Revise A2 Chemistry for OCR Specification A

Contents

Introduction – How to use this revision guide

This revision guide is for the OCR Chemistry A A2 course. It is divided into **units** to match the specification. You may take exams in January and June, or just in June at the end of the course. The content is exactly the same.

In this book, each unit begins with an introduction which summarises the content. It also reminds you of the topics from the AS course which the unit draws on.

The content of each unit is presented in blocks to help you divide your study into manageable chunks. Each block is dealt with in several spreads. These do the following:

- they **summarise** the content
- they indicate **points to note**
- they include **worked examples** of calculations
- they include **diagrams** of the sort you might need to reproduce in exams
- they provide **quick check questions**, to help you test your understanding

At the end of each unit, there are longer questions similar in style to those you will encounter in exams. **Answers** to all questions are provided at the end of the book.

The table below shows you the **scheme of assessment** for the A2 course.

unit name	whole or half unit?	what sort of chemistry?	length of exam	total number of marks in the exam	synoptic content	approximate distribution of marks in the exam		
						structured questions	extended writing questions	quality of written communication
2814 Chains, Rings and Spectroscopy	whole	organic	1 hour 30 mins	90	none	75	25	2
2815/01 Trends and Patterns	half	inorganic	1 hour	45	30 marks	35	10	1
2815/02-06 Option (Biochemistry; Environmental Chemistry; Methods of Analysis and Detection; Gases, Liquids and Solids; Transition Elements)	half	depends on the option chosen	50 mins	45	none	35	10	1
2816/01 Unifying Concepts in Chemistry	half	physical	1 hour 15 mins	60	all	45	15	1
Practical chemistry	half	not covered in this revision guide – your school or college will organise this assessment which is either coursework or a practical exam						

OCR A2 Chemistry

In the A2 year you will complete three units. Two of these units are split into half units, as you can see from the table on the previous page. Just like the AS course, one half unit is practical coursework or exam. However, in the AS course you took three written exams, but there are four written exams in the A2 course. One of these is an Option, and the rest are compulsory.

An important difference between the AS course and the A2 course is the **synoptic content** – there is none in AS, but 40% of the overall marks in A2 are synoptic (this includes practical assessment). The exams where synoptic content is tested is shown in the table. So what does "synoptic" mean? *Synoptic content brings together different areas of chemistry, and shows how they relate to one another.*

In order to answer synoptic questions you will have to know material from the AS course, but don't worry – you won't have to revise the AS units again in detail! But you will have to remember the basic ideas and principles you learnt in the AS course, in much the same way that the material in the AS Foundation unit was needed to answer some questions in the other AS exams.

There are two different kinds of synoptic question:

- the most straightforward type simply asks you about topics from two or more different units. For example, in the Chains, Rings and Spectroscopy exam you may be asked about optical isomers (which you study in the Chains, Rings and Spectroscopy unit) using examples from the alkanes (which you studied in the AS unit Chains and Rings);
- another type of synoptic question asks you about a broad area of chemistry – for example the relationships between different transition metal compounds. You have to show in your answer that you can draw on your knowledge of structure, shapes, bonding and equilibrium *without being told specifically to use these ideas*. In other words, you have shown by yourself how certain different topics are all important in understanding one particular area of chemistry. This type of question usually involves extended writing.

When you are taking an exam you may not realise which marks are synoptic. Don't worry about this – just follow the instructions in the question closely and you will give a synoptic answer if it is required.

Some students worry about the Quality of Written Communication (QWC) marks. In fact, most students are awarded all of these marks. You are always told in which questions the QWC marks are awarded (they will involve some extended writing, which means about five marks for the answer). Sometimes QWC marks are awarded for spelling, punctuation and grammar. Sometimes they are awarded for a well-organised answer which includes relevant scientific or technical terms. You will not know which aspect of QWC is being tested, but if you simply write a good clear account using correct English you will be awarded the marks.

Unit D: Chains, Rings and Spectroscopy

"Chains, Rings and Spectroscopy" extends the carbon chemistry that you learned in "Chains and Rings" and it would be helpful if you reviewed the facts and principles from the previous module.

The table below shows the knowledge from previous modules that will be useful in the revision of this module.

Topic	Reference to specification	Prior knowledge required
Arenes	5.4.1	Foundation: Aspects of covalent bonding. Chains and Rings AS: Structural formulae; π bonding; isomerism; addition and substitution reactions; and electrophiles.
Carbonyl compounds	5.4.2	Foundation: Electronegativity; aspects of covalent bonding including shapes and bond angles. Chains and Ring AS: Structural formulae; π bonding; isomerism and nucleophiles.
Carboxylic acids and esters	5.4.3	Foundation: Molecular shapes and bond angles. Chains and Rings: Structural formulae; How Far How Fast AS: Acids as H^+ donors and properties of acids.
Nitrogen compounds	5.4.4	How Far How Fast AS and Synoptic module: Acids as H^+ donors and acid–base equilibria. Chains and Rings: Structural formulae; isomerism and nucleophiles.
Stereochemistry and organic synthesis	5.4.5	Foundation: Molecular shapes. Chains and Rings: All chemical reactions of the functional groups.
Polymerisation	5.4.6	Chains and rings: Addition and condensation polymerisation.
IR spectroscopy	5.4.7	Chains and rings: Structures of different groups.
NMR spectroscopy	5.4.7	Chains and rings: Structures of different groups.
Mass spectroscopy	5.4.7	Foundation: Mass spectroscopy and relative molecular mass. Chains and rings: Structures of different groups.
Exam questions	5.4.7	Exam sections on chains, rings and spectroscopy.

Arenes

Arene chemistry along with aliphatic (straight chain) chemistry make up the two main branches of organic chemistry. "Arenes" is the "Rings" part of "Chains and Rings"! Arene chemistry is the chemistry of the benzene ring.

Background facts

The molecular formula of benzene is C_6H_6 so its empirical formula is CH. You can spot an arene derivative from the relative numbers of carbon and hydrogen atoms. If the numbers are almost equal or there are more carbons than hydrogens then the compound is an arene, for example methylbenzene is C_7H_8, naphthalene is $C_{10}H_8$.

✓ Quick check 1

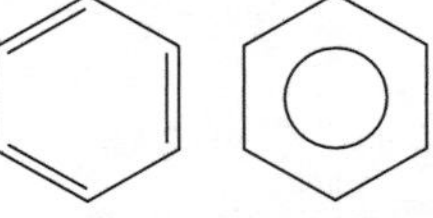
Kekulé structure   Delocalised structure

- The structures of benzene and its related compounds are represented using skeletal formulae. The six C–H bonds in the ring are the corners of the hexagon.

  The skeletal formula of benzene can be written in two different ways.
- The π **electrons** in the C=C double bonds in benzene are **delocalised** (see later) and therefore the delocalised structure is the preferred one of the two.
- Just as alkyl groups are derived from alkanes, for example methyl CH_3– from methane (CH_4), so groups derived from benzene are called **aryl** groups and the presence of a benzene substituent in a compound is indicated by the prefix phenyl-. For example, $C_6H_5CH_2OH$ is phenyl methanol (a hydrogen in methanol is replaced by C_6H_5–).
- Because of the ring structure, benzene compounds can exhibit **positional isomerism**. For example, there are three isomers of dichlorobenzene.

  If the substituents are different, then more than three isomers may be obtained. For example there are four compounds of formula C_7H_7Cl.

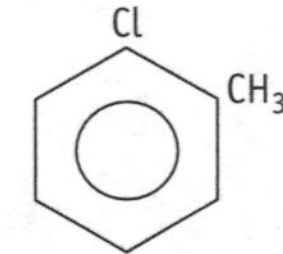

chloro-2-methylbenzene

chloro-3-methylbenzene

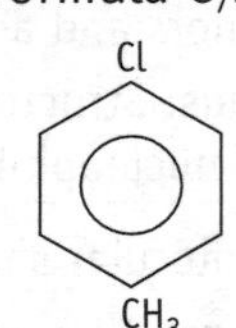

chloro-4-methylbenzene

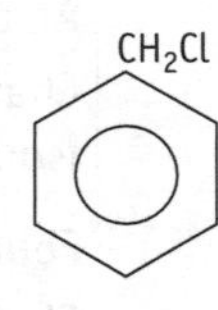

chloromethylbenzene

✓ Quick check 2,3

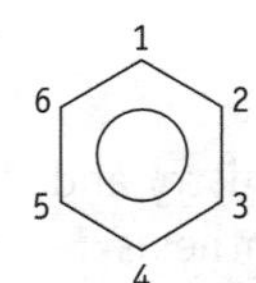

The carbon atoms are numbered 1 to 6 clockwise.

The structure of benzene

Benzene's structure is a very important aspect of its chemistry and should be understood before you learn its reactions. How benzene's structure comes about is shown below:

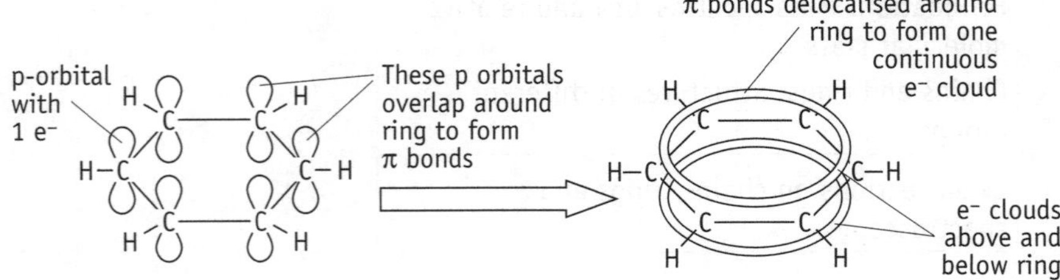

- Benzene is an unsaturated hydrocarbon. However, because the π electrons are delocalised, benzene does not undergo electrophilic addition (as do the alkenes). Rather it undergoes **electrophilic substitution** in which a hydrogen atom in the ring is replaced by another atom or group such as the electrophile NO_2^+.
- Substitution allows the benzene to retain its delocalised system of π electrons and hence its stability.

Remember an **electrophile** is an atom or ion that can accept a pair of electrons.

The reactivity of benzene compared with cyclohexene

Because it also has six carbons, cyclohexene is often compared with benzene because it is a typical alkene.

- The **delocalisation** of the **π electrons** stabilises benzene making it much more unreactive than other unsaturated hydrocarbons, such as the alkenes.
- **The delocalised system of electrons makes benzene more stable than expected if we view it as having three C=C bonds.**

  For example, for benzene to react with bromine, a catalyst is required. The usual catalysts are either iron, iron(III) bromide or aluminium bromide.

$$\textbf{e.g. } C_6H_6 + Br_2 \xrightarrow{FeBr_3} C_6H_5Br + HBr$$

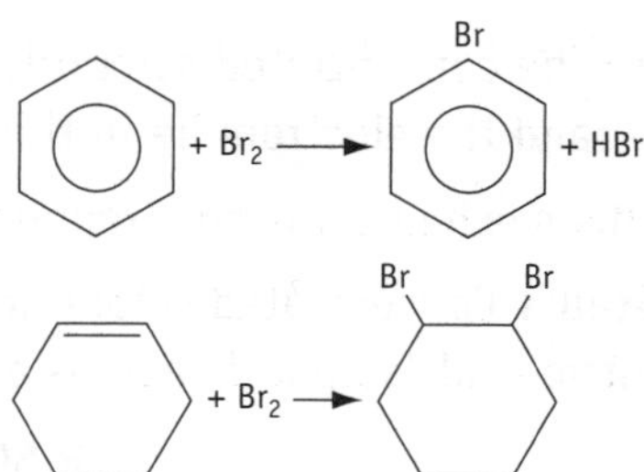

▶▶ *See summary of reactions on page 5.*

This is an electrophilic substitution reaction because the substituting group attacks the electrons in the ring (electrophilic) and replaces one of the hydrogen atoms on the ring (substitution).

- In contrast, the bromination of cyclohexene simply requires bromine water and no catalyst because its π electrons are localised.

$$C_6H_{10} + Br_2 \rightarrow C_6H_{10}Br_2$$

This is an **addition** reaction.

✓ *Quick Check 4*

Quick check questions

1 Which of the following molecular formulae represent arene compounds?
(a) C_7H_8 (b) C_6H_{14} (c) $C_8H_{10}O$ (d) $C_4H_{10}O$

2 Draw the structures of the following compounds:
(a) 1,3-dibromobenzene (b) 1,2-dimethylbenzene
(c) 1,3,5-trichlorobenzene (d) bromo-3-ethylbenzene
(e) chloro-4-methylbenzene

3 Draw the four isomers of the compound with the molecular formula C_8H_{10}.

4 Write equations for the reactions in the dark of chlorine with benzene and the alkene butene ($CH_3CH_2CH{=}CH_2$). Name the different **types** of reaction undergone by the benzene and the alkene and give any conditions (other than light) which are required for the reaction to occur.

Electrophilic substitution

The high electron density above and below the plane of the benzene ring means that the ring is liable to attack from an electron deficient atom or group. Three examples are given.

✓ Quick check 1

Nitration The mononitration of benzene uses concentrated nitric acid and concentrated sulphuric acid. This is the mechanism you should learn.

- The full equation for the reaction is:

$$C_6H_6 + HNO_3 \rightarrow C_6H_5NO_2 + H_2O$$

(benzene) $+ HNO_3 \rightarrow$ (benzene)$-NO_2 + H_2O$

- The concentrated sulphuric acid acts as a catalyst in this reaction and the electrophile in this reaction is the **nitronium ion**.

This mechanism is best represented as a number of steps:

Step 1 Concentrated sulphuric acid acts as an acid and protonates the concentrated nitric acid (sulphuric acid is a stronger acid then nitric acid) which acts as a base.

$$H_2SO_4 + HNO_3 \rightarrow HSO_4^- + H_2NO_3^+$$

The $H_2NO_3^+$ ion then decomposes to give the nitronium ion.

$$H_2NO_3^+ \rightarrow NO_2^+ + H_2O$$

The structure of the nitronium ion is:

Overall we can summarise the reaction as: $H_2SO_4 + HNO_3 \rightarrow HSO_4^- + NO_2^+ + H_2O$

Step 2 The NO_2^+ ion is a powerful electrophile and attracts electrons out of the ring to form a positively charged intermediate.

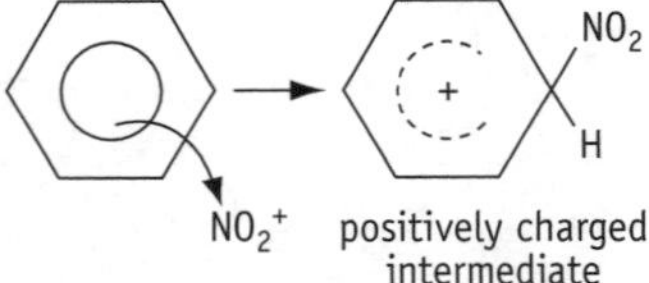

Step 3 The electrons in the positively charged intermediate are only partially delocalised making it unstable. To regain its completely delocalised system a hydrogen ion (proton) is lost to give nitrobenzene.

✓ Quick check 2

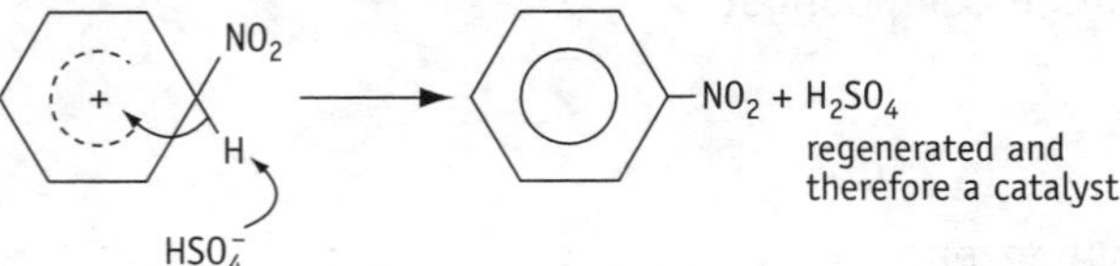

REMEMBER: The curly arrow represents an electron pair

Halogenation Halogen atoms replace hydrogen atoms on the ring. Halogen carriers such as iron, iron(III) bromide or aluminium chloride catalyse the reaction.

(benzene) $+ Br_2 \xrightarrow[FeBr_3]{\text{room temp.}}$ (benzene)$-Br + HBr$

A halogen carrier is an electron-deficient molecule that helps produce an electron-deficient halogen atom or alkyl group (both electrophiles). This can then react with the electrons on the benzene ring.

Alkylation The introduction of an alkyl group such as CH_3– also requires a halogen carrier. This reaction is called the Friedel–Crafts reaction. An example is the preparation of methylbenzene

(benzene) $+ CH_3Cl \xrightarrow[\text{reflux}]{AlCl_3}$ (benzene)$-CH_3 + HCl$

methylbenzene

Summary of the reactions of benzene

✓ Quick check 3

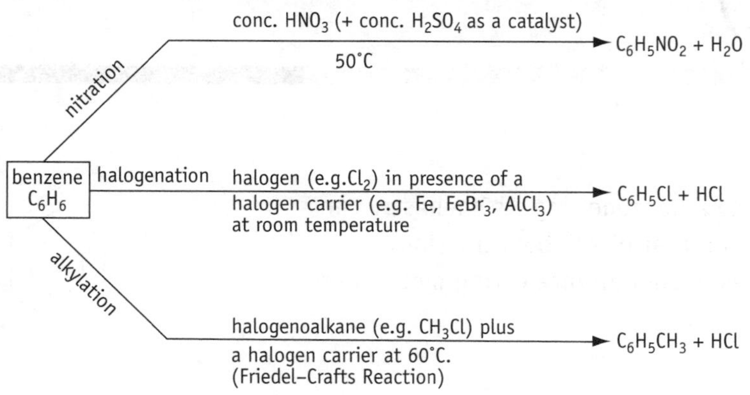

The importance of benzene in industrial synthesis

The above reactions can be extended to form industrially important chemicals.

Examples:

- Formation of the **explosive TNT**

  Benzene can be converted to methylbenzene (toluene) using the Friedel–Crafts reaction.

  The methyl benzene can then be nitrated to give trinitromethylbenzene (**trinitrotoluene** or **TNT**).

- Formation of **polystyrene** (Styrene is phenylethene $C_6H_5CH{=}CH_2$)

  Styrene is formed from benzene by a series of reactions (including alkylation) and then polymerised to give polystyrene.

- The nitration of benzene to nitrobenzene

  Nitrobenzene can be converted to phenylamine which is an important starting point for making **dyes**.

You do not have to know the pathways to the industrially important substances, simply be **aware** that they can be made from benzene.

Quick check questions

1 Explain why benzene undergoes substitution reactions and not addition reactions.

2 Write the balanced equation for the nitration of benzene and write out the mechanism, explaining each step.

3 Complete the following equations and for each one describe the conditions required for the reaction.

(a) $C_6H_6 + Cl_2 \rightarrow$ **A** (organic molecule) + **B**

(b) $C_6H_6 +$ **C** $\rightarrow$ **D** $+ C_6H_5CH_3$

(c) $C_6H_6 + HNO_3 \rightarrow$ **E** (organic molecule) + **F**

(d) $C_6H_6 +$ **G** $\rightarrow$ HBr + **H**(organic molecule)

(e) $C_6H_6 +$ **I** $\rightarrow$ **J** $+ C_8H_{10}$

Phenol (C_6H_5OH)

The structure of phenol is shown opposite.

OH

Because it has a hydroxyl (–OH) group attached to a benzene ring its chemistry can be treated in two parts – that of the –OH group and that of the benzene ring. However, each one affects the other, which makes phenol an interesting molecule to study.

Reactions of the OH group

- Phenol loses the proton on the OH group more easily than alcohols for two reasons.

  i The resulting phenoxide ion ($C_6H_5O^-$) is stabilised by delocalisation of the negative charge on the benzene ring. This means that it is more likely to be formed and therefore this favours losing the H^+ ion.

  ii The lone-pair electrons on the oxygen can be delocalised in the ring and this pulls them into the ring. The net effect of this is to weaken the O–H bond making it easier to lose the H^+ ion, thus making phenol a proton donor.

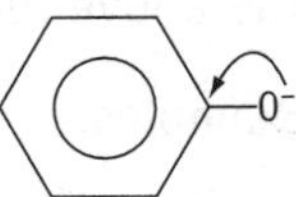

  The overall consequence of both these effects is to give phenol the ability to lose a proton and be a proton donor and hence an acid. Therefore if a proton acceptor or base (Base:) is present the following reaction can take place.

$$C_6H_5OH + Base: \rightarrow C_6H_5O^- + Base:H^+$$

NOTE: The base has a lone pair of electrons which are used to form a dative covalent bond with the proton.

- With bases, the phenol acts as an acid forming water and the phenoxide ion.

  Example: With sodium hydroxide, water and sodium phenoxide are formed.

$$NaOH + C_6H_5OH \rightarrow C_6H_5O^-Na^+ + H_2O$$

Quick check 1

Quick check 2

  The ionic equation can be written:

$$OH^- + C_6H_5OH \rightarrow C_6H_5O^- + H_2O$$

- With sodium, phenol gives sodium phenoxide and hydrogen gas.

$$Na + C_6H_5OH \rightarrow C_6H_5O^-Na^+ + \tfrac{1}{2}H_2$$

Notice the similarity between phenol's reaction with sodium and that of ethanol and other alcohols. The difference is that the alcohols gave an alkoxide ion rather than the phenoxide ion.

Reactions of the benzene ring in phenol

The lone-pair electrons on the oxygen of the –OH group can be delocalised on the benzene ring. This increases the electron density on the ring especially at the sites indicated by the closed circles in the diagram opposite (positions 2, 4 and 6 on the ring) and makes the benzene ring more reactive.

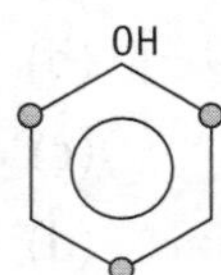

Therefore when the benzene ring of phenol reacts with electrophiles:

- It reacts more readily than benzene because of the increased electron density on the benzene ring. This means that the conditions you need for phenol to react are much less severe than for benzene. The reactants can be more dilute and the temperature will be lower.

- Substitution occurs at the sites indicated in the diagram on the previous page because of the increased electron density at these sites compared with the others.

✓ Quick check 3

- Substitution often occurs at not just one of these sites but at all three because of the extra reactivity.

Example: With bromine:

i No halogen carrier is required. The reaction occurs with bromine water, showing the increased reactivity of the benzene ring.

✓ Quick check 4

ii Substitution occurs at carbons 2, 4 and 6 to give 2,4,6-tribromophenol

$$\mathbf{C_6H_5OH + 3Br_2 \rightarrow C_6H_2Br_3OH + 3HBr}$$

2,4,6-tribromophenol

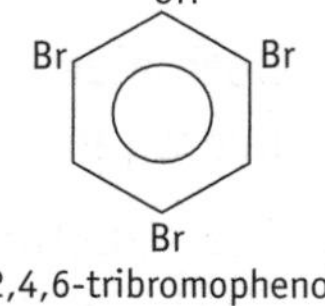

2,4,6-tribromophenol

If chlorine is used instead of bromine then 2,4,6-trichlorophenol (TCP) is formed and this is a valuable antibacterial and antiseptic agent. Another derivative of phenol is sold under the name "Dettol".

Quick check questions

1 Explain why phenol is an acidic compound.

2 Complete the following equations:

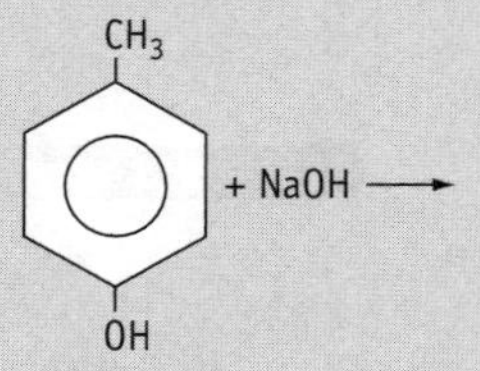

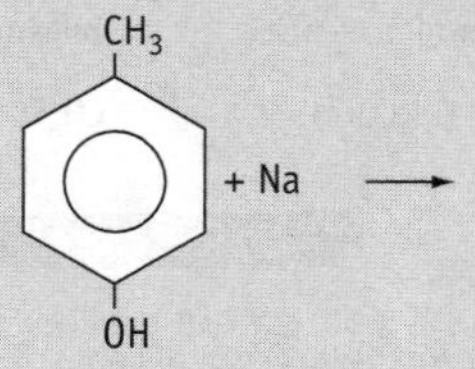

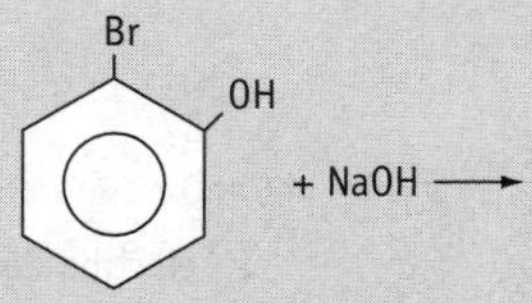

CH_3 / OH (ring) + NaOH →

CH_3 / OH (ring) + Na →

Br / OH (ring) + NaOH →

3 When phenol reacts with bromine water, substitution takes place at the 2, 4 and 6 positions on the benzene ring. Explain why this is so.

4 Give evidence to show that the –OH group in phenol activates the benzene ring.

Carbonyl compounds

The **aldehydes** and **ketones** are two closely related homologous series. They both contain the **carbonyl** (>C=O) group and are therefore called **carbonyl compounds.**

> **NOTE:** Carboxylic acids and esters also contain the >C=O group but have other functional groups on the same carbon atom and can therefore be distinguished from carbonyl

Background facts

- The general formula of carbonyl compounds is $C_nH_{2n}O$ and once we get to three carbons, aldehydes are isomeric with ketones.
- Both aldehydes and ketones contain less than the maximum possible number of hydrogens and are therefore **unsaturated.**
- The structural difference between aldehydes and ketones is that aldehydes have a hydrogen attached to the carbonyl carbon, whilst ketones do not. However, because of the carbonyl group both will give an intense absorption in the 1680–1750 cm^{-1} region. The NMR spectra of aldehydes show a peak with a large chemical shift around 10 ppm because of the proximity of the aldehyde proton to the electronegative oxygen.
- When naming aldehydes the names end in –al. Examples of aldehydes are ethanal and butanal.
- The names of ketones end in –one, and a number preceding it indicates where on the carbon chain the carbonyl group is situated. Two examples are propan-2-one and butan-2-one.
- The bond angles in aldehydes and ketones are all close to 120° and they are trigonal planar in shape.

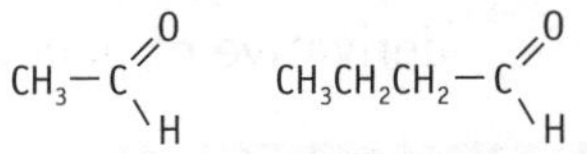

	ethanal	butanal
	$CH_3{-}C(=O)H$	$CH_3CH_2CH_2{-}C(=O)H$
structural formula	CH_3CHO	$CH_3CH_2CH_2CHO$

Quick check 1

	propan-2-one	butan-2-one
	$(H_3C)(H_3C)C{=}O$	$(H_3C)(CH_3CH_2)C{=}O$
structural formula	CH_3COCH_3	$CH_3COCH_2CH_3$

Quick check 2

Nucleophilic addition to carbonyl compounds

The electronegative oxygen adjacent to the carbonyl carbon means that a dipole is produced in the C=O bond and the carbon is electron deficient.

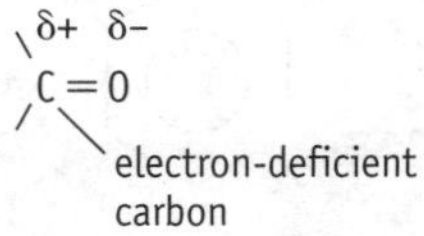

$$\overset{\delta+}{C}{=}\overset{\delta-}{O}$$

electron-deficient carbon

The carbonyl carbon can therefore be attacked by nucleophiles.

Because the compounds are unsaturated the reaction is also an addition, so the reaction mechanism is that of **nucleophilic addition.** The example shows the mechanism for the reaction between ethanal and hydrogen cyanide.

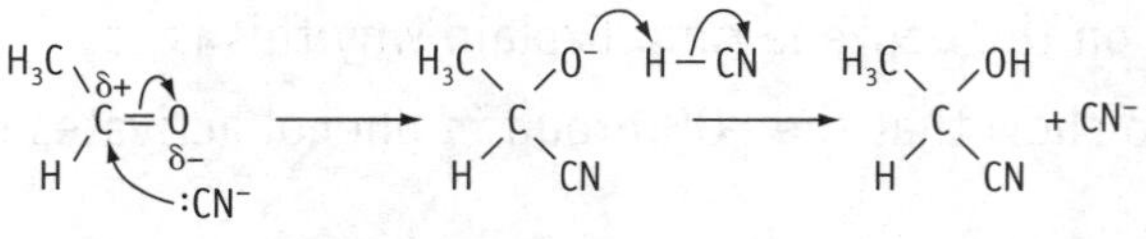

$$H_3C(H)\overset{\delta+}{C}{=}\overset{\delta-}{O} + :CN^- \longrightarrow H_3C(H)C(CN){-}O^- \xrightarrow{H-CN} H_3C(H)C(CN)OH + CN^-$$

Step 1 Nucleophilic attack by CN^- ion on carbonyl carbon

Step 2 Reaction of intermediate ion with proton from HCN

Final product 2-hydroxy-propanenitrile

Generally, the reaction between any carbonyl compound and HCN may be written as shown.

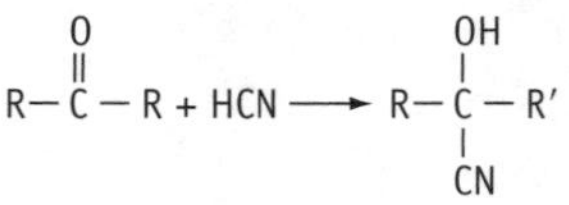

$$R{-}\overset{O}{\overset{\|}{C}}{-}R + HCN \longrightarrow R{-}C(OH)(CN){-}R'$$

R and R′ can be hydrogen, an alkyl group (e.g. $-CH_3$) or a phenyl group (e.g. $-C_6H_5$).

This reaction is particularly important for aldehydes because the length of the carbon chain is increased by one carbon. The product is called a nitrile because of the –C≡N group. For example, CH_3CN is ethanenitrile.

> REMEMBER: **Nucleophiles** are compounds or ions that can donate a pair of electrons. In this example it is the CN^- ion. In practice KCN is used (this gives a higher concentration of CN^- ions) along with a low concentration of acid (to supply H^+ ions).

The reduction of carbonyl compounds

Quick check 4

The reagent, sodium borohydride in water ($NaBH_4$) can be used to reduce (by adding hydrogen) **polar unsaturated** bonds such as the C═O bond in aldehydes and ketones.

Group	Example
Aldehydes are reduced to primary acohols **$RCHO + 2[H] \rightarrow RCH_2OH$**	Ethanal is reduced to ethanol **$CH_3CHO + 2[H] \rightarrow CH_3CH_2OH$**
Ketones are reduced to secondary alcohols **$RCOR' + 2[H] \rightarrow RCH(OH)R'$**	Butan-2-one is reduced to butan-2-ol **$CH_3COCH_2CH_3 + 2[H] \rightarrow CH_3CH(OH)CH_2CH_3$**

Distinguishing and identifying carbonyl compounds

You can distinguish between aldehydes and ketones using **ammoniacal silver nitrate** solution and then identify which aldehyde or ketone is present using **2,4-dinitrophenylhydrazine**.

	Aldehydes	Ketones
Distinguishing test Add ammoniacal silver nitrate solution. **(a mild oxidising agent)**	They are oxidised to a carboxylic acid and a **silver mirror** is formed. **$RCHO + [O] \rightarrow RCOOH$**	No reaction
Distinguishing test Add acidified potassium dichromate solution.	They are oxidised to carboxylic acid and the potassium dichromate changes colour from orange to green. **$RCHO + [O] \rightarrow RCOOH$**	No reaction and therefore the dichromate stays orange in colour.
Identification test – Add 2,4-dinitrophenylhydrazine solution. **This also confirms the presence of the carbonyl group**	They give an orange crystalline precipitate called a hydrazone. Each aldehyde gives a hydrazone with a characteristic melting point after it has been recrystallised and therefore can be identified.	They give an orange crystalline precipitate called a hydrazone. Each ketone gives a hydrazone with a characteristic melting point after it has been recrystallised and therefore can be identified.

The silver ions are reduced because they gain electrons, $Ag^+ + e^- \rightarrow Ag$

NOTE: [O] represents the oxygen from the oxidising agent.

The structure of 2,4-dinitrophenyl hydrazine is:

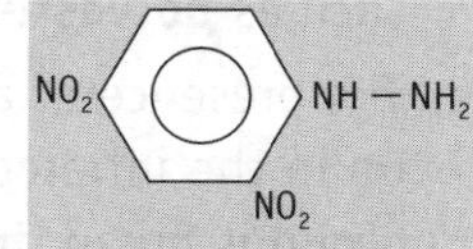

Quick check questions

1 Name the following aldehydes:

(a) $CH_3CH_2CH_2CHO$; (b) $CH_3CH(CH_3)CHO$; (c) C_6H_5CHO; (d) $C_6H_5CH_2CHO$

2 Name the following ketones:

(a) CH_3COCH_3; (b) $CH_3CH_2COCH_2CH_3$; (c) $C_6H_5CH_2COCH_3$; (d) $CH_3COCH_2CH_2C_6H_5$

3 Complete the following equations giving the structure of the carbon compound formed.

(a) $CH_3CHO + HCN \rightarrow$

(b) $CH_3COCH_2CH_3 + HCN \rightarrow$

4 Name and give the structural formulae of the products after reducing the following carbonyl compounds: (a) CH_3CHO; (b) $CH_3CH_2COCH_3$; (c) $C_6H_5COCH_3$; (d) $CH_3CH(CH_3)CHO$

Carboxylic acids and esters

Carboxylic acids are the strongest acids you will study in A-level organic chemistry. They are important because they are the starting materials for making esters and they are found naturally, making them readily available.

Carboxylic acids

Background facts

- The functional group present is the **carboxyl** group –COOH group which is trigonal planar in shape with bond angles of 120°.
- Their general formula is $C_nH_{2n}O_2$. Their names are derived from the parent alkanes.

Name	Molecular formula	Structural formula
Methanoic acid	CH_2O_2	HCOOH
Ethanoic acid	$C_2H_4O_2$	CH_3COOH
Butanoic acid	$C_4H_8O_2$	$CH_3CH_2CH_2COOH$

✓ Quick check 1

- If there are two carboxyl groups in the molecule, then again the name is derived from the parent alkane and the two carboxyl groups indicated by the suffix –dicarboxylic acid (or dioic acid).

HO–C(=O)–CH₂–C(=O)–OH

propan-1,2-dicarboxylic acid or propan-1,2-dioic acid

HO–C(=O)–C₆H₄–C(=O)–OH

benzene-1,4-dicarboxylic acid

Dicarboxylic acids are important in the preparation of **condensation polymers** such as **polyesters** which are used for making fibres for clothing.

- The presence of a carbonyl (>C=O) group and the hydroxyl (–OH) group shows up in the infrared spectra of carboxylic acids. The absorption due to the –OH group is broad, indicating hydrogen bonding.
- The hydrogen on the –COOH group is highly electron deficient because of the proximity of the two electronegative oxygens. The NMR spectra of carboxylic acids illustrate this by the large chemical shift of the –COOH proton at values for δ greater than 10.

Ethanoic acid molecules form hydrogen bonds as shown below.

--- hydrogen bond

Carboxylic acids as acids

- In the carboxyl group the two electronegative oxygens pull electrons towards themselves and away from the hydrogen in the –OH group. This weakens the O–H bond (see below), making it easier to lose the hydrogen as a H^+ ion (proton).
- The carboxylate ion ($-COO^-$) is stabilised by the delocalisation of the negative charge on the ion easing its formation (see below).

✓ Quick check 2

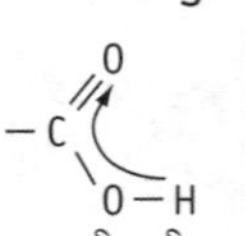
electrons pulled in this direction away from the hydrogen, thus weakening the O–H bond

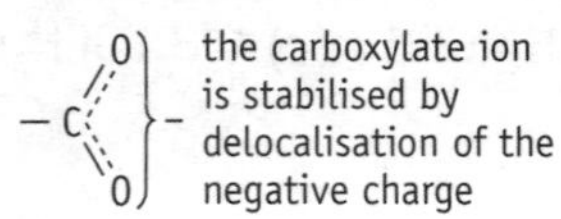
the carboxylate ion is stabilised by delocalisation of the negative charge

- In aqueous solution water acts as a base (proton acceptor), and solutions of carboxylic acids have pH values of around 4–5, making them weak acids. Therefore the equilibria in the reaction below lies towards the left-hand side.

  Example: ethanoic acid:

  $$\mathbf{CH_3COOH + H_2O \rightleftharpoons CH_3COO^- + H_3O^+}$$
  $$\textbf{acid} + \textbf{base} \rightleftharpoons \textbf{conjugate base} + \textbf{conjugate acid}$$

- With alkalis such as sodium hydroxide they react just like other acids in forming a salt plus water and are therefore neutralised by the alkali.

  $$\mathbf{CH_3COOH + NaOH \rightarrow CH_3COO^-Na^+ + H_2O}$$
  ethanoic acid → **sodium ethanoate**

  The ionic equation may be written:

  $$\mathbf{CH_3COOH + OH^- \rightarrow CH_3COO^- + H_2O}$$
  ethanoic acid → **ethanoate ion**

✓ Quick check 3

Esterification

In the presence of **acid catalysts** and heat, carboxylic acids and alcohols react to form **esters and water**. The reaction is a **reversible** one.

$$\text{carboxylic acid + alcohol} \underset{\text{heat}}{\overset{\text{acid (H}^+\text{) catalyst}}{\rightleftharpoons}} \text{ester + water}$$

reversible reaction

Example: Ethanoic acid and ethanol react to form ethyl ethanoate and water.

$$\mathbf{CH_3COOH(l) + C_2H_5OH(l) \rightleftharpoons CH_3COOC_2H_5(l) + H_2O(l)}$$

- The acid catalyst may be either concentrated hydrochloric or sulphuric acid. It is the hydrogen ions supplied by the acids which are the actual catalysts.
- If concentrated sulphuric acid is used, not only does it supply hydrogen ions for catalysis but it also absorbs water. If it absorbs water then the concentration of water on the right-hand side of the equilibrium is lowered, disturbing the equilibrium. According to **Le Chatelier's principle**, in order to restore the equilibrium, the concentration of water has to be raised and to do this *the equilibrium shifts to the right*, not only raising the concentration of water again but at the same time raising the concentration of the ester.

Esterification may be looked upon as a **condensation** reaction. In this reaction type, two small molecules join together to form a larger molecule (in this case the ester) with the elimination of a small molecule (in this case water). In **condensation polymerisation** this happens many times over to form a **polyester** (see later section).

✓ Quick check 4

Quick check questions

1. Give the structural formulae of the following acids:
   (a) propanoic acid (b) 2-methylpropanoic acid (c) 2-phenylethanoic acid
2. (a) Explain why carboxylic acids are able to donate protons.
   (b) Predict which of 2-chloroethanoic acid ($CH_2ClCOOH$) or ethanoic acid (CH_3COOH) is the stronger acid and give a reason for your answer. NOTE: chlorine is an electronegative element.
3. Complete the following equations:
   (a) $CH_3COOH + NaOH$
   (b) $CH_3CHClCOOH + OH^-$
   (c) $C_6H_5COOH + OH^-$
   (d) $CH_3CH(CH_3)COOH + NaOH$
4. Explain why concentrated sulphuric acid might be preferred to concentrated hydrochloric acid as the catalyst in esterification.

Esters

Esters are found naturally and are worth studying because examiners like to give questions on them.

Facts you should know

- Esters are isomeric with carboxylic acids; therefore, their general formula is $C_nH_{2n}O_2$. They can however, be distinguished from carboxylic acids by their infrared spectra *which do not show* the broad absorption band at approximately 3000 cm^{-1}.

✓ Quick check 1

- Their characteristic functional group is the ester bond (–CO–O–). Structurally they may be represented as in the diagram below:

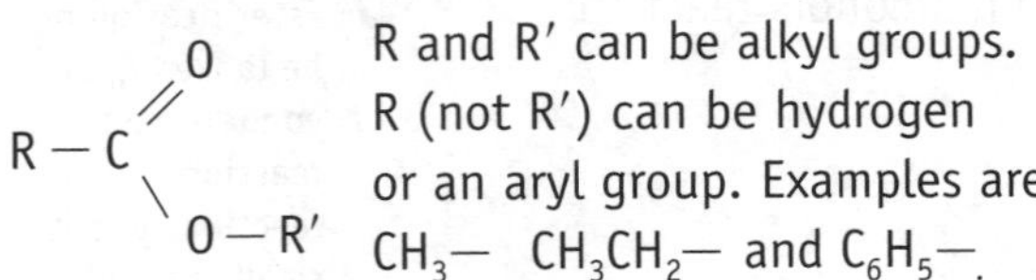

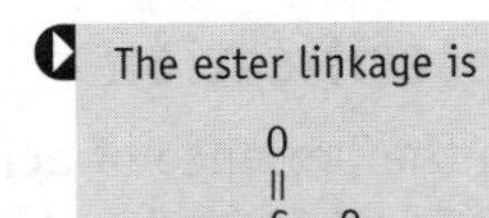

- Esters are named according to the acid and alcohol from which they are prepared. The alcohol part of the name comes first and the acid part second. For example, ethyl ethanoate and methyl propanoate:

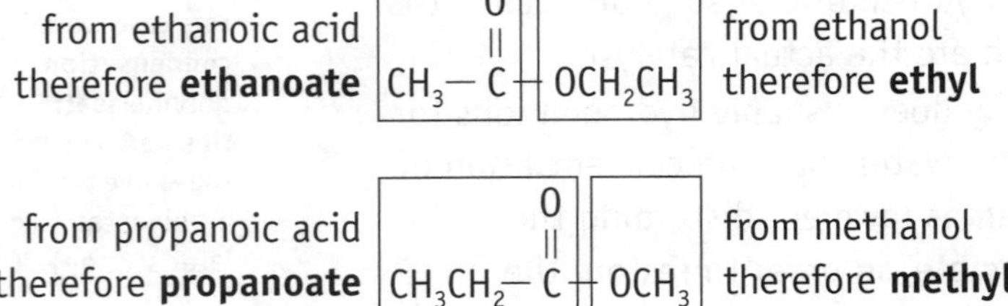

Strange isn't it – naming them back to front like this? However those are the rules and you have got to know how to do it!

Uses of esters

- Esters have pleasant smells and because of their reasonably low boiling points are volatile and evaporate easily. Consequently, they are used in perfumes and flavourings. Many natural perfumes are esters, for example oil of wintergreen and the smells given from flowers. They are often used in cheap artificial scents. In flavourings, examples of esters used are ethyl methanoate (raspberry essence) and pentyl ethanoate (pear essence).
- Esters are good solvents and used for dissolving such substances as drugs and antibiotics.
- Waxes are esters of long chain carboxylic acids and long chain alcohols. Fats are esters of long chain carboxylic acids and the alcohol, glycerol (propan-1,2,3-triol, $CH_2OHCHOHCH_2OH$). When fats are hydrolysed by alkalis (see later), salts of long chain fatty acids (soaps) are formed and the process is called saponification.

✓ Quick check 3

Hydrolysis of esters

- When esters are refluxed with aqueous acids (as **catalysts**) the reverse of the esterification reaction occurs and the acid and alcohol are obtained. The reaction is reversible. It is a hydrolysis reaction because water is split (hydro = water; lysis = splitting) as shown below:

$$R-\overset{\displaystyle O}{\overset{\|}{C}}-O-R' + HO-H \xrightarrow{\text{hydrolysis}} R-\overset{\displaystyle O}{\overset{\|}{C}}-OH + HO-R'$$

acid alcohol

For example, ethyl ethanoate is **hydrolysed** to ethanoic acid and ethanol.

$$\mathbf{CH_3COOC_2H_5 \;+\; H_2O \;\rightleftharpoons\; CH_3COOH \;+\; C_2H_5OH}$$

ester **water** **acid** **alcohol**

✓ Quick check 4

A fat would be hydrolysed to glycerol and the carboxylic acids combined with it.

- If a base (e.g. NaOH), is used as a catalyst, then the first stage produces an alcohol and an acid as with acid catalysts, but the acid produced reacts further with the base in a neutralisation reaction.

For example, with ethyl ethanoate we again get ethanoic acid plus ethanol. The ethanoic acid then reacts with OH^- ions from the base to give ethanoate ions plus water. This reaction goes to completion.

Stage 1: $CH_3COO\ C_2H_5 + H_2O \rightleftharpoons CH_3COOH + C_2H_5OH$

Stage 2: $CH_3COOH + OH^- \rightarrow CH_3COO^- + H_2O$

Simplified mechanisms for both forms of hydrolysis are shown opposite.

acid hydrolysis

$R-C(=O)-OR' \xrightarrow{H^+} R-C^+(OH)-OR' \xrightarrow[\text{rearrangement}]{\text{attack by water on } C^+ \text{ then}} R-C(OH)(OH)-OR' \xrightarrow{H^+} R-C(=O)OH + R'OH$

alkaline hydrolysis

$R-\overset{\delta+}{C}(=O)-OR' + OH^- \longrightarrow R-C(=O)OH + R'O^-$ $\quad R'O^- + H_2O \longrightarrow R'OH + OH^-$

Quick check questions

1 Give the structural formulae of all the isomers (carboxylic acids and esters) with the following molecular formulae:
(a) $C_3H_6O_2$ (3 isomers) (b) $C_4H_8O_2$ (5 isomers)

2 (a) Give the structural formulae and the names of the carboxylic acids forming the following esters:
(i) $CH_3CH_2COOCH_3$ (ii) $HCOOCH_3$ (iii) $CH_3CH_2CH_2COOCH_2CH_3$
(b) Name the esters in (i) to (iii) above.

3 Name some uses for esters and explain why you think they have these uses.

4 Complete the following equations:
(a) $CH_3COOC_2H_5 + H_2O \rightleftharpoons$
(b) $C_6H_5COOCH_3 + H_2O \rightleftharpoons$
(c) $CH_3CH_2COOC_6H_5 + H_2O \rightleftharpoons$
(d) $HCOOCH_2CH_2CH_3 + H_2O \rightleftharpoons$

Nitrogen compounds

These form a group of very important compounds. Primary amines are industrially important because of their uses in making dyes. Amino acids are biologically extremely important.

Primary amines

Background facts

- They are derivatives of ammonia (NH_3)
- The functional group is the amino $-NH_2$ group.

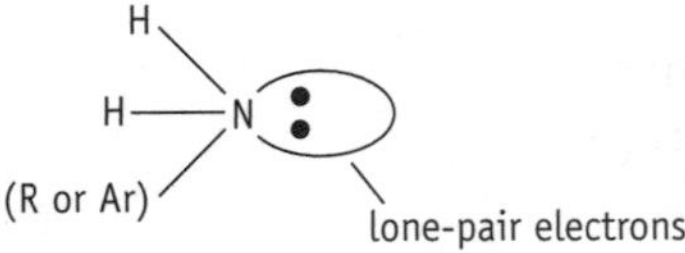

R = alkyl group (chain)
Ar = aryl group (benzene ring)

- Their general formula is $C_nH_{2n+1}NH_2$ and they are named according to which group they are attached to.

  Examples are given in the table below:

Structural formula	Name
CH_3NH_2	Methylamine
$CH_3CH_2NH_2$	Ethylamine
$C_6H_5NH_2$	Phenylamine

R is an alkyl group whilst Ar is an aryl group (derivative of benzene)

✓ Quick check 1

The basic nature of primary amines

- The lone pair on the nitrogen of primary amines means that they can form dative covalent bonds with protons (H^+ ions). The new ($-NH_3^+$) ion is derived from an ammonium ion and is named accordingly. Examples are:

$$CH_3CH_2NH_2 + H^+ \rightarrow CH_3CH_2NH_3^+$$

ethylamine → **methylammonium ion**

$$C_6H_5NH_2 + H^+ \rightarrow C_6H_5NH_3^+$$

phenylamine → **phenylammonium ion**

Because they are lone-pair donors, amines can be classified as **nucleophiles.**

- Because they are proton acceptors, primary amines are Bronsted–Lowry bases.
- With water, they lead to the formation of OH^- ions and this makes them weak alkalis.

$$\text{e.g. } CH_3CH_2NH_2 + H_2O \rightleftharpoons CH_3CH_2NH_3^+ + OH^-$$

- **The basicity of the amino (—NH_2) group** depends on the electron density of the lone-pair electrons on the nitrogen atom. The greater the electron density the more likely they are to accept a proton and the more basic the amine.

✓ Quick check 2

The H^+ ions need two electrons to be stable and these are provided by a dative covalent bond with the nitrogen.

i Alkyl groups have a **positive inductive** effect because they **push electrons away** from themselves. Therefore, they would increase the electron density on the nitrogen atom.

ii A benzene ring would allow the lone-pair electrons to be **delocalised into the ring** and this would decrease the electron density on the nitrogen atom. Therefore, aromatic amines are less likely to capture a proton.

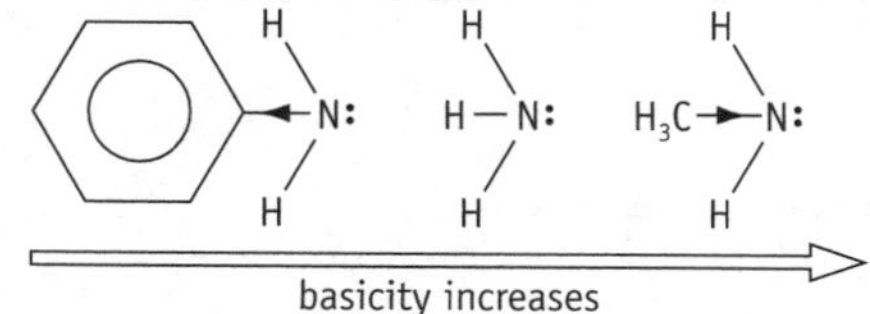

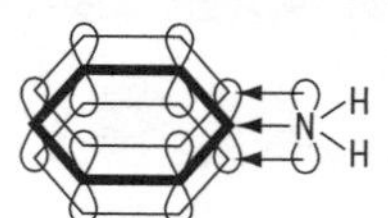

For phenylamine, the lone-pair electrons on the nitrogen can interact with the π-orbitals on the ring and are **delocalised**.

Alkyl groups are electron-donating groups

✓ *Quick check 3*

Phenyl groups e.g. C_6H_5 are electron-withdrawing groups.

- Because they act as bases, primary amines react with acids to form salts. These salts are derivatives of ammonium salts. The derivative of ammonia is named by taking the ammonium ion, NH_4^+, and inserting the name of the substituting group. In the two examples given, an ethyl group (CH_3CH_2–) and a phenyl group (C_6H_5–) substitute for a hydrogen so the salts are named ethylammonium chloride and phenylammonium chloride.

For example:

$$\mathbf{NH_3 + HCl \rightarrow NH_4^+Cl^-}$$
ammonia — **ammonium chloride**

$$\mathbf{CH_3CH_2NH_2 + HCl \rightarrow CH_3CH_2NH_3^+Cl^-}$$
ethylamine — **ethylammonium chloride**

$$\mathbf{C_6H_5NH_2 + HCl \rightarrow C_6H_5NH_3Cl^-}$$
phenylamine — **phenylammonium chloride**

✓ *Quick check 4*

? *Quick check questions*

1 Name the following compounds:
(a) CH_3NH_2 (b) $CH_3CH_2CH_2NH_2$ (c) $CH_3CH(CH_3)CH_2NH_2$ (d) $C_6H_5NH_2$

2 (a) Complete the following equations:
(i) $CH_3NH_2 + H_2O \rightleftharpoons$
(ii) $C_6H_5NH_2 + H_2O \rightleftharpoons$
(b) How are the amines in (a) able to act as bases?

3 (a) Place the following compounds in order of basicity (least basic first). Explain your answer.
NH_3; CH_3NH_2; $C_6H_5NH_2$
(b) The nitro group ($-NO_2$) is an electron-withdrawing group. Explain whether 4-nitrophenylamine (see figure opposite for structure) is more or less basic than phenylamine.

4 Complete the following equations and name the products:
(a) $CH_3NH_2 + HCl \rightarrow$
(b) $CH_3CH_2CH_2NH_2 + HCl \rightarrow$
(c) $C_6H_5NH_2 + HCl \rightarrow$

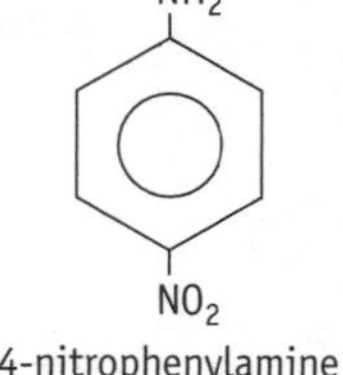

4-nitrophenylamine

The chemistry of phenylamine ($C_6H_5NH_2$)

Preparation

- Phenylamine is prepared by the reduction of nitrobenzene using a mixture of tin and concentrated hydrochloric acid.
- The reaction mixture is refluxed to ensure the reaction is complete.

$$C_6H_5NO_2 + 6[H] \xrightarrow[\text{from tin and conc. HCl}]{\text{reflux}} C_6H_5NH_2 + 2H_2O$$

nitrobenzene

- Phenylamine is the starting compound for the formation of an important class of compounds – the azo dyes – and the reaction scheme and accompanying details for this are given below.

✓ Quick check 1

NOTE: Aliphatic (straight chain) amines such as ethylamine are prepared by the reduction of nitriles (R-CN) using sodium and ethanol as the source of hydrogen.
$R\text{-}CN + 4[H] \rightarrow RCH_2NH_2$

Phenylamine and azo-compounds

Stage 1: Sodium nitrite ($NaNO_2$) and hydrochloric acid react to form nitrous acid. The temperature should be below 10 °C because the benzenediazonium chloride decomposes at higher temperatures.

Stage 2: When the benzenediazonium chloride is then added to phenol in the presence of an alkali, an azo dye is formed.

The equations showing the structures and conditions are given below along with the full reaction scheme.

✓ Quick check 2

below 10°C **Stage 1**

$$C_6H_5NH_2 + 2HCl + NaNO_2 \longrightarrow C_6H_5N_2^+Cl^- + NaCl + H_2O$$

benzenediazonium chloride

Stage 2

$$C_6H_5N_2^+Cl^- + C_6H_5OH \xrightarrow[\text{of alkali}]{\text{in presence}} C_6H_5\text{-}N{=}N\text{-}C_6H_4OH$$

benzenediazonium chloride; phenol; 4-(phenylazo)phenol

Full reaction scheme

$$C_6H_5NH_2 \xrightarrow[\text{Stage 1}]{NaNO_2/HCl \text{ below } 10°C} C_6H_5N_2^+Cl^- \xrightarrow[\text{Stage 2}]{C_6H_5OH \text{ in alkali}} C_6H_5\text{-}N{=}N\text{-}C_6H_4OH$$

phenylamine; benzenediazonium chloride; 4-(phenylazo)phenol

Economically this reaction is very important since the formation of azo-dye compounds is the basis of part of the dye industry. A common example of an azo dye is the acid–base indicator, methyl orange.

Amino acids

Background facts

Their general formula is $RC^*H(NH_2)COOH$ and their general displayed formula can be represented as shown opposite.

NH_2
R—C*—COOH
H

The asterisk (*) indicates the chiral carbon (see text)

If R is not hydrogen, the carbon atom (marked with an asterisk) is attached to four different atoms/groups. It is therefore a **chiral** carbon, and amino acids can exhibit **optical isomerism**. Optical isomers can be drawn as mirror images of each other.

▸▸ *More about optical isomerism on p. 20.*

R, C, COOH, NH_2, H — HOOC, C, R, NH_2, H

mirror images

The acid–base nature of amino acids

All amino acids contain the following two groups:

1 The basic amine ($-NH_2$) group. This group can accept protons.
2 The acidic carboxyl (–COOH) group. This group can donate protons.

When you draw optical isomers, you should try to give some idea of three dimensions to the figures.

- Because they contain these two groups, they can react with both acids and bases and are therefore **amphoteric**.
- As acids, the carboxyl group can donate a proton to a base to form a salt plus water.

$$RCH(NH_2)COOH + NaOH \rightarrow RCH(NH_2)COO^-Na^+ + H_2O$$

- As bases, the amino group can accept a proton from an acid to form a salt.

$$RCH(COOH)NH_2 + HCl \rightarrow RCH(COOH)NH_3^+Cl^-$$

✓ Quick check 3

Quick check questions

1 (a) Complete the following equations and name the organic product:

(i) $C_6H_5NO_2 + 6[H] \rightarrow$ (ii) $C_6H_4(CH_3)NO_2 + 6[H] \rightarrow$
4-methylnitrobenzene

(b) What reagents are used to produce the hydrogen in the above reactions?

2 (a) Complete the following reaction schemes:

(i) $C_6H_5NH_2 + NaNO_2 + 2HCl \rightarrow$ **A** $+ H_2O + NaCl$ **A** $+ C_6H_5OH \rightarrow$ **B**

(ii) $C_6H_4(CH_3)NH_2 + NaNO_2 + 2HCl \rightarrow$ **C** $+ H_2O + NaCl$
2-methylphenylamine **C** $+ C_6H_4(CH_3)OH \rightarrow$ **D** 32-methylphenol

(b) What is the economic importance of the compounds **B** and **D**?

3 (a) Complete the following equations:

(i) $CH_2(NH_2)COOH + HCl \rightarrow$ (ii) $CH_2(NH_2)COOH + NaOH \rightarrow$

(b) Explain why the amino acid is acting as a base in reaction (i) and as an acid in reaction (ii)

Amino acids (zwitterions), proteins and peptides

Amino acids are white crystalline solids which have much higher melting points than would be expected from their molecular masses. The explanation for these properties is that at intermediate pH values they exist as a type of "inner salt" or **zwitterion** where the charge on the amino group ($-NH_3^+$) cancels out that on the carboxylate group ($-COO^-$). The general form of the zwitterion is shown in the diagram.

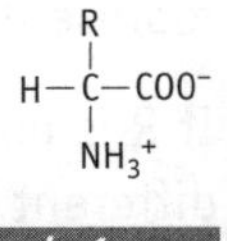

$$H-\underset{NH_3^+}{\overset{R}{C}}-COO^-$$

✓ Quick check 1

- At high pH values the proton on the $-NH_3^+$ will be removed leaving an overall negative charge.
- At low pH values the $-COO^-$ group will accept a proton giving an overall positive charge.
- At an intermediate pH called the isoelectric point, there will be equal negative and positive charges and the amino acid will have no overall charge.

✓ Quick check 2

The positive and negative charges on each zwitterion attract opposite charges on neighbouring zwitterions, forming ionic bonds. These are stronger than the hydrogen bonds that would exist between the uncharged amino acids. This stronger ionic bonding explains the higher than expected melting points of amino acids.

✓ Quick check 3

Proteins and peptides

Facts you should know

- Proteins are chains of amino acids linked by peptide bonds. The peptide bond is the CO-NH link.
- Proteins are formed from amino acids by the **loss of water** as the peptide bonds are formed. Proteins are therefore **condensation polymers.**

A polypeptide is one containing more than ten amino acids – usually 100–300

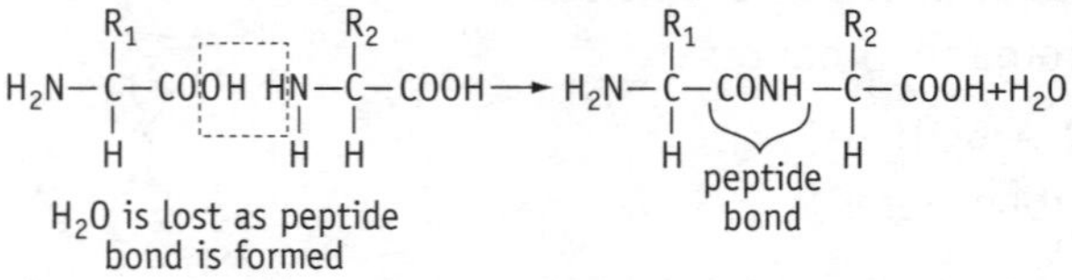

$$H_2N-\underset{H}{\overset{R_1}{C}}-COOH + HN(H)-\underset{H}{\overset{R_2}{C}}-COOH \longrightarrow H_2N-\underset{H}{\overset{R_1}{C}}-CONH-\underset{H}{\overset{R_2}{C}}-COOH + H_2O$$

H_2O is lost as peptide bond is formed

peptide bond

A common exam question involves the structures of the **di**peptide formed when **two** amino acids combine. The point is that the amino acids can combine in two different ways to give compounds with **different** structures.

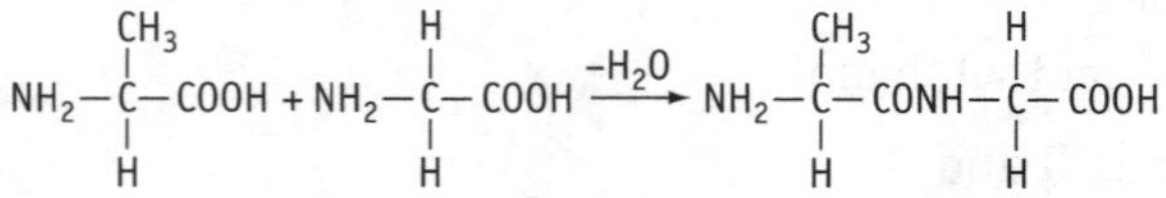

$$NH_2-\underset{H}{\overset{CH_3}{C}}-COOH + NH_2-\underset{H}{\overset{H}{C}}-COOH \xrightarrow{-H_2O} NH_2-\underset{H}{\overset{CH_3}{C}}-CONH-\underset{H}{\overset{H}{C}}-COOH$$

In this reaction the carboxyl group of 2-aminopropanoic acid reacts with the amine group of 2-aminoethanoic acid.

$$NH_2-\underset{H}{\overset{H}{C}}-COOH + NH_2-\underset{H}{\overset{CH_3}{C}}-COOH \xrightarrow{-H_2O} NH_2-\underset{H}{\overset{H}{C}}-CONH-\underset{H}{\overset{CH_3}{C}}-COOH$$

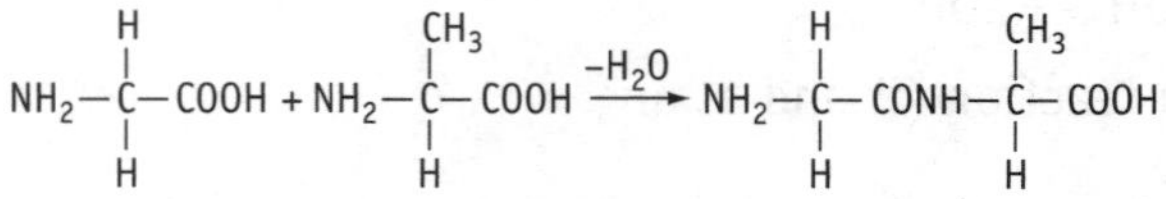

In this reaction the amine group of 2-aminopropanoic acid reacts with the carboxyl group of 2-aminoethanoic acid.

- To get the individual amino acids back from a polypeptide, water has to be added back to the polypeptide and the peptide bonds broken. This is another example of hydrolysis.
- When a protein or peptide is refluxed with concentrated (approximately 6 mol dm^{-3}) hydrochloric acid for several hours, it is hydrolysed and the protein is split up into its constituent amino acids. This is essentially the reverse of the formation of the peptide bond.

$$-HN-\underset{H}{\overset{R_1}{C}}-CO-NH-\underset{H}{\overset{R_2}{C}}-CO\cdots \quad (HO-H) \xrightarrow{HCl(aq)} -HN-\underset{H}{\overset{R_1}{C}}-COOH + H_2N-\underset{H}{\overset{R_2}{C}}-CO\cdots$$

Peptide bond is hydrolysed as it reacts with water

✓ Quick check 4

Quick check questions

1 Draw the zwitterions formed from the following amino acids:

(a) $CH_2(NH_2)COOH$ (b) $CH_3CH(NH_2)COOH$ (c) $C_6H_5CH(NH_2)COOH$

2 The isoelectric points of the acids shown above are as follows:

2-aminoethanoic acid	2-aminopropanoic acid	2-phenyl-2-aminoethanoic acid
$CH_2(NH_2)COOH$	$CH_3CH(NH_2)COOH$	$C_6H_5CH(NH_2)COOH$
pH 5.97	pH 6.00	pH 5.48

(a) Draw the ions that would be formed by 2-aminoethanoic acid at:

(i) pH 2.00 (ii) pH 5.97 (iii) pH 10.00

(b) Draw the structures of the predominant ions that would be formed by:

(i) 2-aminopropanoic acid at pH 5.75

(ii) 2-phenyl-2-aminoethanoic acid at pH 5.75

3 Why are structures of the predominant the melting points of the three amino acids in question 2 greater than would be expected from their molecular masses?

4 (a) Draw the dipeptides that would be formed by combining the amino acids below:

(i) $CH_2(NH_2)COOH$ and $CH_3CH(NH_2)COOH$

(ii) $CH_2(NH_2)COOH$ and $C_6H_5CH(NH_2)COOH$

(iii) Two molecules of $CH_2(NH_2)COOH$

(b) Draw the structures of the amino acids responsible for the formation of the dipeptides shown below:

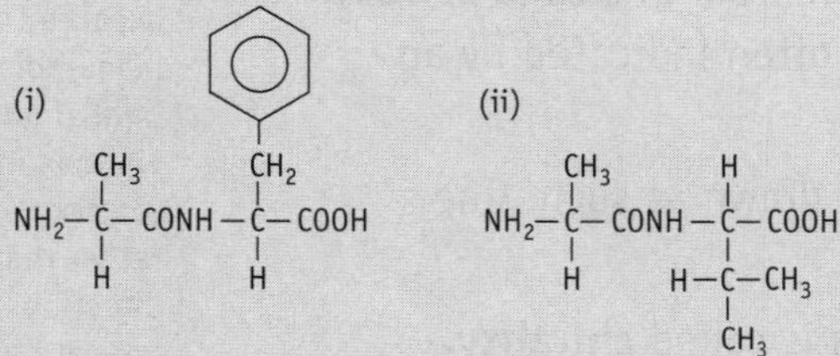

Stereoisomerism and organic synthesis

Stereoisomerism is of paramount importance to the organic chemist especially if the compound is to be used in living systems. There are two types of stereoisomerism – *cis-trans* (geometric) isomerism and optical isomerism. An important natural example of *cis-trans* isomerism is the existence of *cis* and *trans* isomers of retinol (vitamin A) in the mechanism of vision. Vast numbers of enzymes and molecular receptors are also able to recognise optical isomers and if the isomer is the wrong one it will not be used. For example, the smell of lemons and oranges come from the same compound, limonene, but from different optical isomers of limonene.

Stereoisomerism

Stereoisomers have the same structural formula but different displayed formulae. That is, they differ by the arrangement of their atoms in space. As stated above, there are two types, *cis-trans* isomerism and optical isomerism.

Cis-trans isomerism

- This requires a **C=C double bond** (as found in alkenes) and **two different groups on each carbon atom in the double bond.**
- *Cis* isomers have similar groups on both sides of the double bond whilst *trans* isomers have them on opposite sides.

  Examples are the two isomers of but-2-ene.

```
H      CH3          H      H
 \    /              \    /
  C = C               C = C
 /    \              /    \
CH3    H           CH3    CH3
```

trans-but-2-ene (groups on opposite sides of C = C bond)

cis-but-2-ene (groups on same sides of C = C bond)

✓ Quick check 1

Optical isomerism

- This requires a **chiral centre** or **chiral carbon** (one carbon atom **attached to four different groups or atoms**). Each **chiral** carbon atom is often indicated by an asterisk*.
- **Optical isomers** are mirror images of each other and are drawn as such. One isomer cannot be superimposed on the other.
- The property where a compound exists as optical isomers is called **chirality.**

NOTE: You are not required to know the following, but you will see the facts in textbooks. Optical isomers are called enantiomers. They can be distinguished from one another because they rotate plane-polarised light (light that vibrates in one direction only rather than in all directions, as with normal light). The isomer rotating the light clockwise is called the (+) isomer, whilst the other is the (-) isomer.

Examples are amino acids and 2-hydroxypropanoic acid

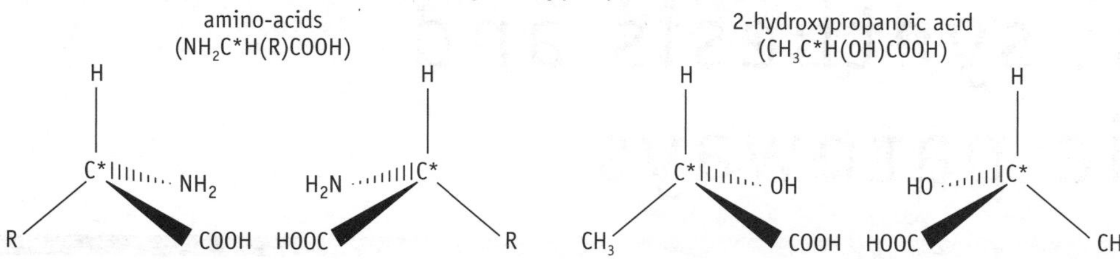

✓ Quick check 2

Organic synthesis and chirality

A compound containing a chiral carbon can be synthesised from a simpler molecule by means of a **synthetic pathway** involving one or more chemical steps. 2-hydroxypropanoic acid (lactic acid) can be synthesised from the simpler compound, ethanal, as follows:

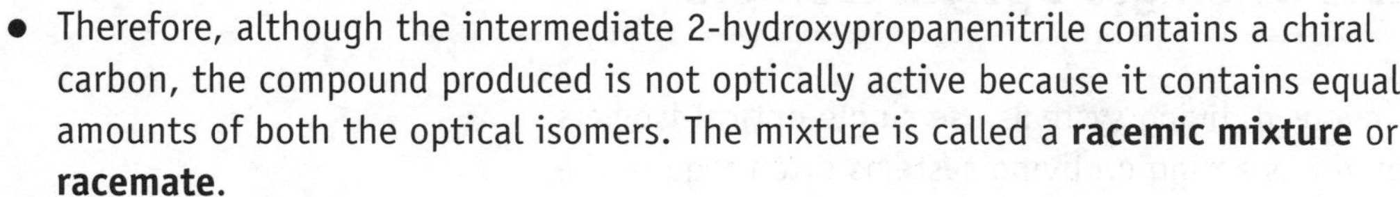

ethanal 2-hydroxypropanenitrile 2-hydroxypropanoic acid

- For aldehydes, the addition of –CN **increases the length of the carbon chain by one** (e.g. in the example above the length goes up from two carbon atoms in ethanal to three carbon atoms in 2-hydroxypropanenitrile).
- The addition of the cyanide ion (CN^-) to the ethanal is a **nucleophilic addition** and because the ethanal molecule is planar, nucleophilic attack on the electron-deficient carbonyl carbon can come from either above or below the plane of the molecule (see diagram opposite).
- Therefore, although the intermediate 2-hydroxypropanenitrile contains a chiral carbon, the compound produced is not optically active because it contains equal amounts of both the optical isomers. The mixture is called a **racemic mixture** or **racemate.**
- Consequently the final product, 2-hydroxypropanoic acid, is also optically inactive even though it has a chiral carbon atom. This outcome is common in the synthesis of chiral molecules in the laboratory especially when planar molecules are used in the process.

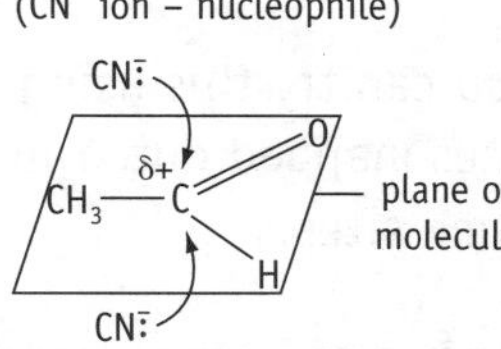

✓ Quick check 3

Quick check questions

1 (a) Draw the isomers of the following alkenes:
 (i) but-2-ene (ii) $C_6H_5CH{=}CHCH_3$
 (b) Why doesn't but-1-ene exhibit *cis-trans* isomerism?

2 (a) Draw the optical isomers of $CH_3CH(OH)COOH$. In your diagram indicate the chiral carbon atoms with an asterisk.
 (b) Explain why $CH_3C(CH_3)(OH)COOH$ is not optically active.

3 When $CH_3CH(OH)COOH$ is synthesised from ethanal, the resulting compound is optically inactive. Explain why.

Chemical synthesis and synthetic pathways

In many ways this section brings together a lot of the understanding and knowledge you have accumulated in the AS sections from last year and those in this module. Quite often the problems posed by having a final product and a starting material from which to synthesise it are ideal ways of learning your organic reactions and conditions and therefore we suggest you learn the flowchart opposite.

Natural synthesis of chiral molecules

- In living systems chiral molecules exist in the majority of cases as one optical isomer only.
- The reason for this "handedness" is that these molecules are synthesised using enzymes, and enzymes are catalysts that can recognise optical isomers. The active site on an enzyme itself possesses a "handedness" and can recognise the four different groups in space. The other optical isomer will not fit these sites and is not used by the enzyme.

You can try this using models and placing them on some paper with the different sites mapped out. You will find that only one of the optical isomers will fit on the same sites.

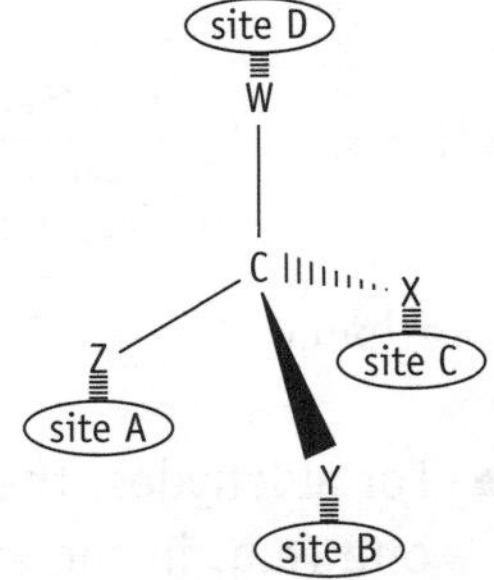

Sites A to D on the enzyme can recognise the spatial orientations of the groups W to Z on the molecule

✓ Quick check 1

Laboratory synthesis of single optical isomers

When chiral molecules are involved, living systems use single optical isomers. Therefore pharmaceutical products acting on living systems often require the synthesis of these single optical isomers. As well as the effectiveness of the drug or chemical (e.g. perfume) other factors have to be taken into account. These are summarised in the table below:

✓ Quick check 2

FACTOR	EXPLANATION	EXAMPLE
Effectiveness	One isomer could exert a much stronger effect than the other could. Therefore, if it is used on its own it will be more effective in lower doses.	One optical isomer of the drug L-DOPA is much more effective than the other. Administering the more effective isomer on its own is less wasteful and causes fewer side-effects.
Toxicity	One isomer could have a beneficial effect whilst its mirror image could well be toxic.	The drug thalidomide is a chiral compound. One optical isomer stops morning sickness in pregnant women, the other causes deformities in unborn foetuses.

Organic synthesis

The interrelationships between the different functional groups are summarised in the flowchart below. This flowchart is a useful summary of much of the carbon chemistry you have learned on the AS and A2 Chemistry courses.

✓ Quick check 3, 4

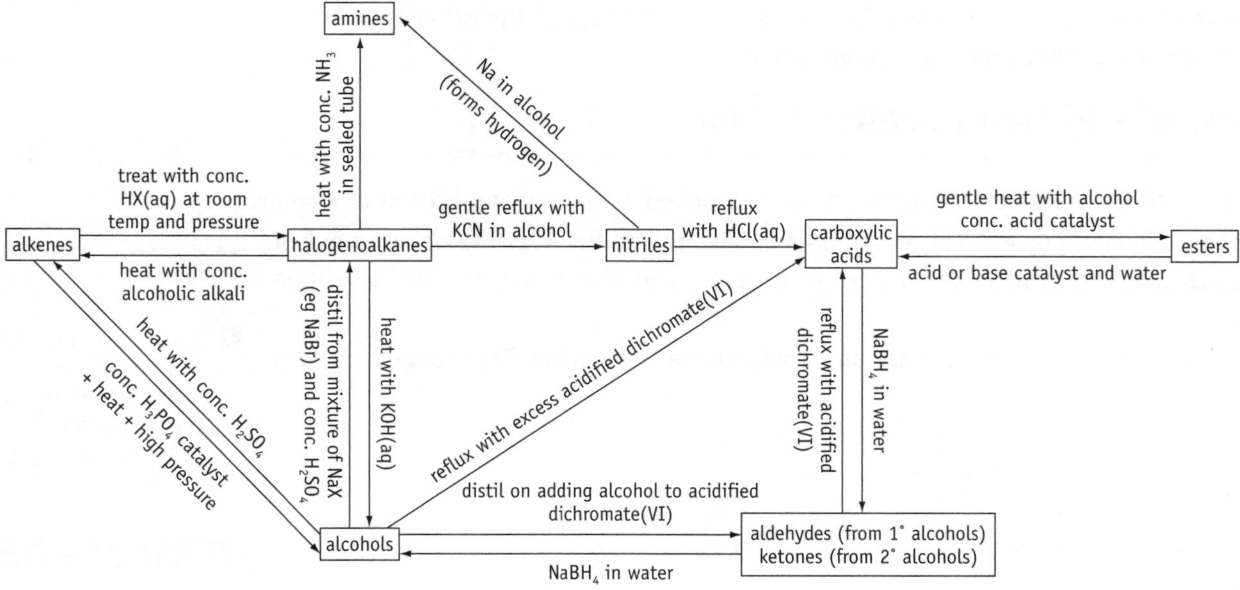

Quick check questions

1 How is it that natural systems are able to synthesise pure optical isomers of different compounds?

2 Give **two** reasons why it is important to synthesise the correct optical isomer when preparing a drug.

3 Explain how the following conversions could be carried out.

(a) $CH_2{=}CH_2 \rightarrow CH_3CH_2OH \rightarrow CH_3COOH \rightarrow CH_3COOC_2H_5$

(b) $CH_3OH \rightarrow CH_3Br \rightarrow CH_3CN \rightarrow CH_3COOH$

(c) $CH_3CH_2\ CH_2CHO \rightarrow CH_3CH_2CH_2CH(OH)CN \rightarrow CH_3CH_2CH_2CH(OH)COOH \rightarrow CH_3CH_2CH_2CHBrCOOH$

4 For the following conversions, identify the possible missing compounds. The conditions are given on the arrows.

$$CH_3CHO \xrightarrow{\text{HCN and gentle reflux}} \mathbf{A} \xrightarrow{\text{HCl (aq) and reflux}} \mathbf{B} \xrightarrow{\text{heat with NaBr and conc. } H_2SO_4} \mathbf{C}$$

$$CH_3CHO \xrightarrow{\text{heat with } NH_3 \text{ in sealed tube}} \mathbf{D}$$

Polymerisation

You have already studied addition polymerisation as part of the chemistry of the alkenes at AS level. We will review this and then go on to look at the other type of polymerisation – **condensation polymerisation**.

Review of addition polymerisation

In the AS section on Addition Polymerisation we looked at the polymerisation of alkenes.

Remember, addition polymerisation requires an alkene and involves the opening out of the double bonds to connect the molecules together and give a polymer with all single bonds.

The diagram below of the polymerisation of polyethene summarises the changes involved.

n represents a large number and the brackets indicate the repeating unit in the polymer.

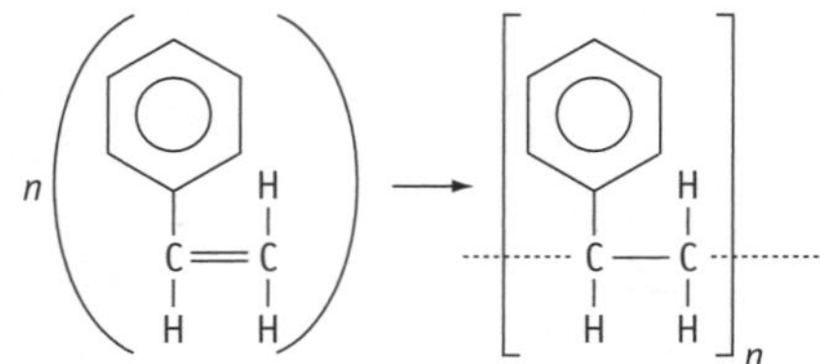

✓ Quick check 1

As well as forming long chains the polymers can be made in such a way as to get crystalline or soft, elastic, rubbery materials. For example, polypropene exists in three forms, each with a distinctive three-dimensional structure. The table below summarises their structures and properties.

Polymer type	Structure	Properties
Isotactic	The methyl groups are on the same position in each repeating unit.	Highly crystalline, since the regular structure allows the chains to fit together.
Syndiotactic	The methyl groups alternate regularly from side to side.	Again, highly crystalline, since the regular structure allows the chains to fit together.
Atactic	The methyl groups are arranged at random.	The chains cannot fit together regularly so the solid is soft, elastic and rubbery.

The skeletal formulae for these three polymer types are shown below:

CH_3 CH_3 CH_3 CH_3
isotactic

CH_3 CH_3 CH_3 CH_3
syndiotactic

CH_3 CH_3 CH_3 CH_3
atactic

✓ Quick check 2

Condensation polymerisation

- A condensation reaction may be defined as one in which two molecules join together to form a larger molecule with the elimination of a smaller molecule.

X–□–X + Y–○–Y ⟶ XY + X–□–○–Y

simple molecule   larger molecule

- In condensation polymerisation the requirements are two compounds each with two functional groups. This means that the particular reaction can take place twice on the same molecule. A general example is given opposite:

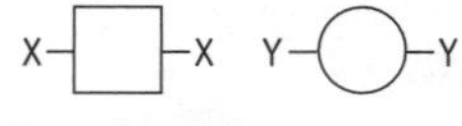

Examples of the groups X and Y are —OH, —COOH and $—NH_2$

- The reaction still leaves X and Y available for bonding and the reaction can continue to give a polymer. In this general case:

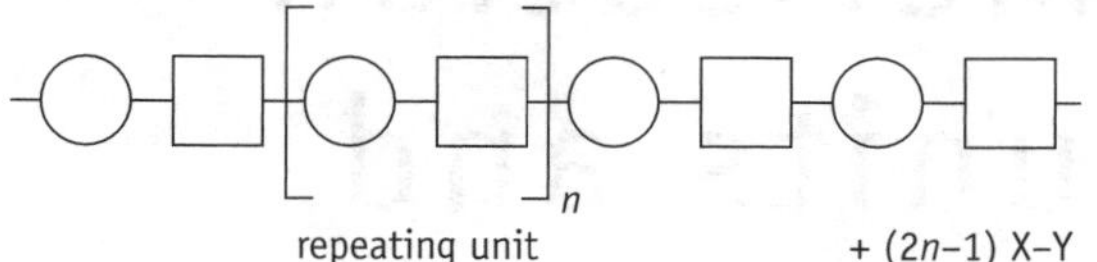

- The lines represent bonds between the units and these are usually either the ester (–CO–O–) bond or the amide (–CO–NH) bond.

Examples are shown in the table below:

No.	X–□–X	Y–○–Y	Molecule (XY)	Bond formed, type of polymer and repeating unit (shown by brackets)
1.	Ethane-1,2-diol $HOCH_2CH_2OH$	Benzene-1,4-dicarboxylic acid HO–C(=O)–C_6H_4–C(=O)–OH	H_2O	Ester (–CO–O–) polyester [–OCH_2CH_2O–C(=O)–C_6H_4–C(=O)–]
2.	1,6-diamino-hexane $H_2N(CH_2)_6NH_2$	Hexane-1,6-dicarboxylic acid $HO_2C(CH_2)_4CO_2H$	H_2O	Amide (CO–NH–) polyamide [–C(=O)–$(CH_2)_4$–C(=O)–$NH(CH_2)_6NH$–] Nylon 6,6
3.	Benzene-1,4-diamine H_2N–C_6H_4–NH_2	Benzene-1,4-dicarboxylic acid HO–C(=O)–C_6H_4–C(=O)–OH	H_2O	Amide (CO–NH–) polyamide [–C(=O)–C_6H_4–C(=O)–NH–C_6H_4–NH–] KEVLAR
4.	Amino acid[1] H_2N–C(R')(H)–C(=O)–OH	Amino acid[2] H_2N–C(R'')(H)–C(=O)–OH	H_2O	Amide (peptide) bond (CO–NH–) polypeptide [–HN–C(R')(H)–C(=O)–][–NH–C(R'')(H)–C(=O)–] each amino acid is a unit

Quick check questions

✓ Quick check 3

1 Draw the polymers formed from the monomer units (a) to (d) below.

(a) C(CN)(CN)=C(CN)(CN)   (b) C(CH_3)(CH_3)=C(H)(H)   (c) C(H)(H)=C(Cl)(H)   (d) C(F)(F)=C(F)(F)

2 Explain the differences between isotactic, syndiotactic and atactic polypropene.

3 Ethan-1,2-diol ($HOCH_2CH_2OH$) can form condensation polymers. Draw the repeating units of the polymers it forms with the compounds shown below.

(a) HO–C(=O)–C(=O)–OH   (b) HO–C(=O)–C_6H_4–C(=O)–OH

Condensation polymerisation – monomers, repeat units and uses of condensation polymers

A common task set by examiners is the identification of repeat units and the monomers that are their components. This section is designed to give you some idea of how to tackle this type of question.

Deducing the repeat units and monomers in condensation polymers

A typical condensation polymer is shown below.

To determine the repeat unit for the polymer, go through the following steps:

Step 1: Draw a bracket line as shown, where a distinctive group begins or ends. In this case we have selected between the carbonyl and amine groups.

Step 2: Follow the chain until you come to just before this is about to repeat itself and draw the other bracket.

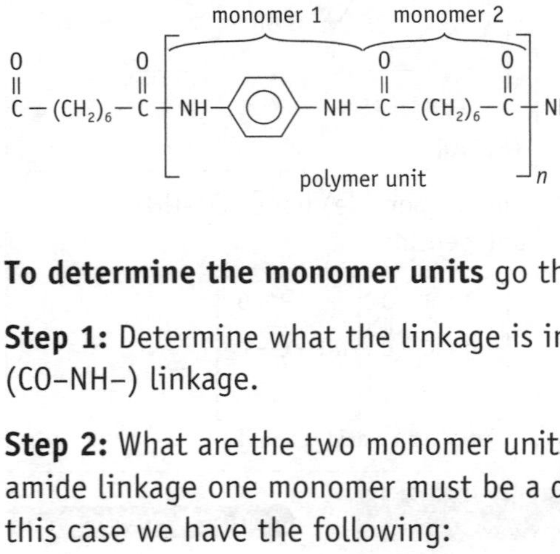

monomer 1 monomer 2

$-C(=O)-(CH_2)_6-C(=O)-[-NH-C_6H_4-NH-C(=O)-(CH_2)_6-C(=O)-]_n-NH-C_6H_4-NH-$

polymer unit

Quick check 1

To determine the monomer units go through the following steps:

Step 1: Determine what the linkage is in the polymer. In this case it is the amide (CO–NH–) linkage.

Step 2: What are the two monomer units on either side of the linkage? To make an amide linkage one monomer must be a diamine and the other a dicarboxylic acid. In this case we have the following:

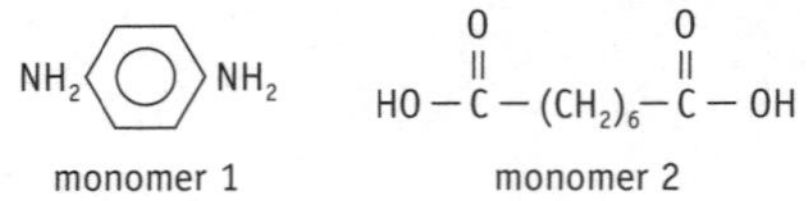

$NH_2-C_6H_4-NH_2$ (monomer 1)

$HO-C(=O)-(CH_2)_6-C(=O)-OH$ (monomer 2)

Quick check 2

Nylons and how they are named

One form of nylon is the polyamide formed from the dicarboxylic acid, hexane-1,6-dicarboxylic acid and the diamine, 1,6-diaminohexane.

The repeating unit for the polymer is shown opposite:

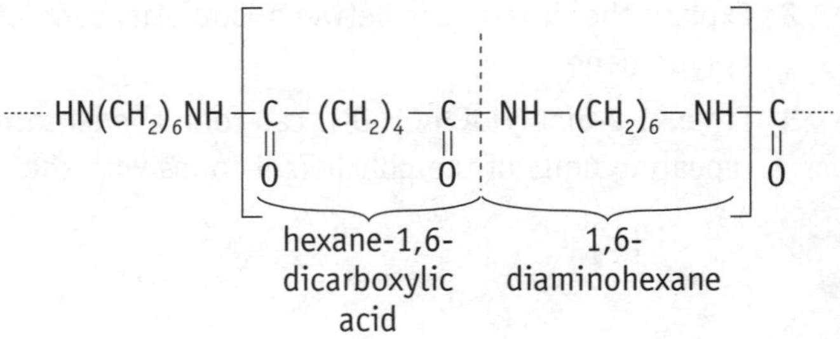

$\cdots HN(CH_2)_6NH-[-C(=O)-(CH_2)_4-C(=O)-NH-(CH_2)_6-NH-]-C(=O)\cdots$

hexane-1,6-dicarboxylic acid | 1,6-diaminohexane

The repeating unit is formed from two compounds each with six carbons and because of this the nylon is called nylon-6,6.

The uses of polyesters and polyamides in fibres

- The strong bonds in polyamides and polyesters make the polymer chains themselves strong and ideal for making fibres.
- When polyamides such as nylon are drawn the molecular chains are aligned along the length of the fibre. Hydrogen bonds can form between the chains and when the fibre is stretched and then released these bonds re-form to bring the molecules back to their original positions making the nylon elastic. This is why nylon is good for making tights.
- For polyesters such as Terylene, cross-linking occurs through dipole–dipole bonding. Again, the fibres produced are elastic and therefore good for making clothing.

  The two inter-chain interactions are shown below:

✓ Quick check 3

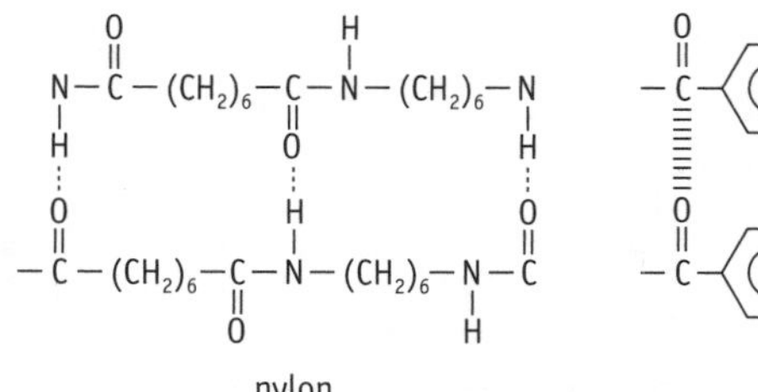

nylon

dipole–dipole

polyester

- In Kevlar (used for bullet-proof vests), the perfect alignment of the chains, and multiple hydrogen bonding, make it a very strong material.

Quick check questions

1 Identify the repeat units and the linkages formed in the polymers (a) to (c) below.

(a) $\cdots OCH_2CH_2{-}OC(=O){-}CH_2CH_2C(=O){-}OCH_2CH_2OC(=O)CH_2CH_2OC(=O)CH_2CH_2O\cdots$

(b) $\cdots C(=O){-}(CH_2)_4{-}C(=O){-}O{-}(CH_2)_6{-}O{-}C(=O){-}(CH_2)_4{-}C(=O){-}O{-}(CH_2)_6{-}O{-}C(=O){-}(CH_2)_4{-}C(=O)\cdots$

(c) $\cdots C(=O){-}C_6H_4{-}C(=O){-}NH{-}C_6H_4{-}NH{-}C(=O){-}C_6H_4{-}C(=O){-}NH{-}C_6H_4{-}NH{-}C(=O){-}C_6H_4{-}C(=O)\cdots$

2 For each polymer in question 1 identify the **two** monomer units present.

3 (a) Explain why nylon is elastic.

(b) Why are nylons and polyesters good fibres?

Infrared spectroscopy

Infrared spectroscopy is a very useful technique in the determination of a compound's structure. It relies on the fact that different bonds in compounds absorb at different frequencies in the infrared region of the spectrum. In infrared analysis the **wavenumbers** (the reciprocal of wavelength) are used instead of frequencies or wavelengths and are measured in cm^{-1}. The functional groups and bonds you will be asked about are given in the table below along with their wavenumbers.

Note that you do not have to remember the wavenumbers in the exam, as they will be given in your databook.

Functional group/bond	Wavenumber (cm^{-1})
Hydroxyl (-OH) in carboxylic acids	Very broad 2500 to 3500
O–H hydrogen bonded in alcohols and phenols	Less broad between 3230 to 3550
Carbonyl group(C=O) in aldehydes, ketones, carboxylic acids and esters	1680-1750
C-O bond in alcohols, carboxylic acids and esters	Between 1000 and 1250

Some typical spectra for compounds exhibiting these infrared absorptions are shown on the opposite page.

Note the following:

- The **alcohol** shows a strong broad absorption at about 3400 cm^{-1} due to O–H hydrogen bonding and no absorption at 1740 cm^{-1} because there is no carbonyl group (C=O).
- The **carboxylic acid** shows a very broad absorption at around 2500–3500 cm^{-1} because of the of the O–H and at approximately 1700 cm^{-1} due to the carbonyl group.
- The **aldehyde** (and a **ketone)** shows an absorption at around 1700 cm^{-1} but no strong broad absorption at ~3000 cm^{-1}.
- The **ester** absorbs at 1700 cm^{-1} but there is no broad absorption at approximately 3000 cm^{-1}. The ester can therefore be distinguished from the isomeric carboxylic acid.
- For the alcohols, esters and carboxylic acids there is a C-O absorption between 1000 and 1300 cm^{-1}.
- Esters and carbonyl compounds both absorb at ~1700 cm^{-1} and show no absorption for the O-H group. The carbonyl does not, however, show any absorption for the C-O bond. They can also be easily distinguished using other means. For example, their molecular formulae are different as are their chemical reactions.

– OH alcohols

–C(=O)–O–H carboxyl group

>C=O carbonyl compounds (aldehydes and ketones)

–C(=O)–O–R esters

Quick check1,2&3

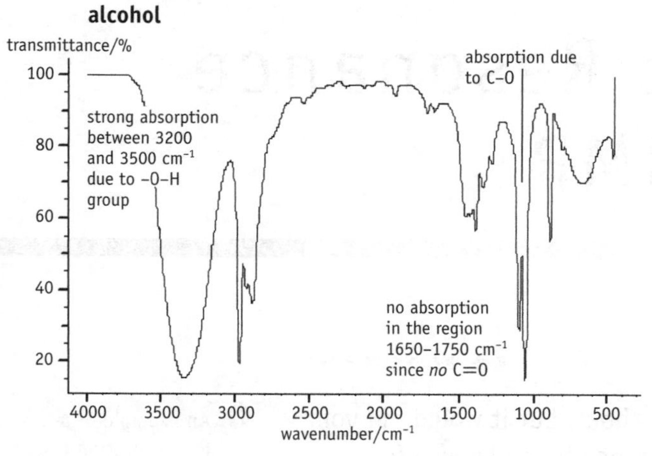

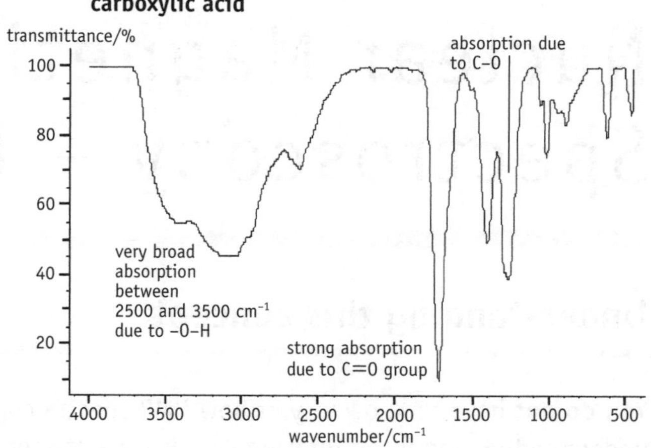

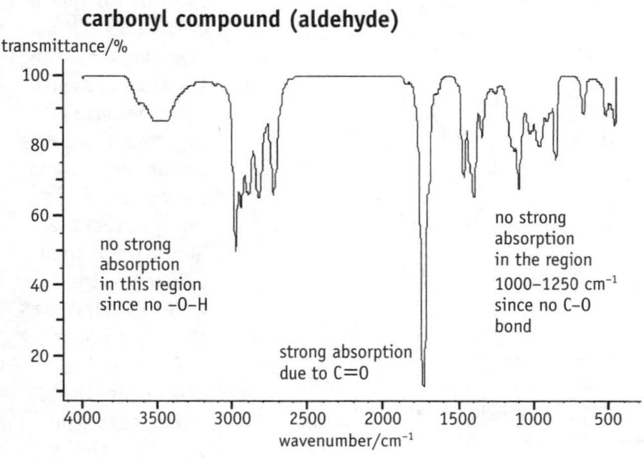

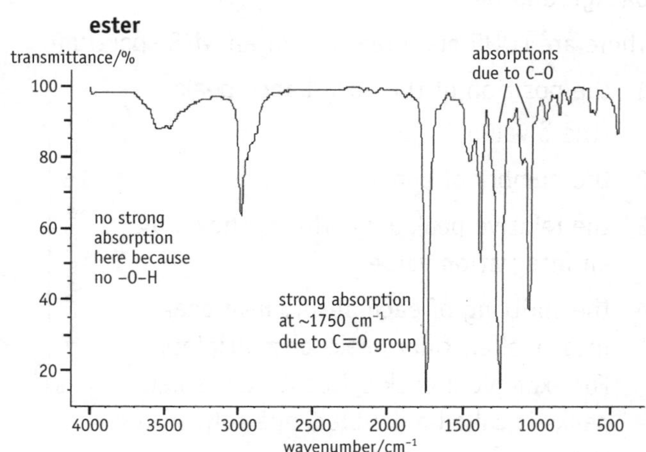

Quick check questions

1 The compound $C_4H_8O_2$ has a number of isomers. Two of them are given below.

$CH_3COOCH_2CH_3$ $CH_3CH_2CH_2COOH$

(a) Which of them would give a strong sharp absorption at 1700 cm^{-1} as well as a broad absorption at ~3000 cm^{-1}.

(b) Explain your answer.

2 A compound, X, gives a strong, broad absorption at around 3000 cm^{-1} but none at 1700 cm^{-1}.

(a) Which functional group is present in X?

(b) To which homologous series does it belong? Explain your reasoning.

3 Describe what changes you would see in the infrared spectra when ethanol (CH_3CH_2OH) is oxidised to ethanal (CH_3CHO).

Nuclear Magnetic Resonance Spectroscopy – NMR

Understanding this concept

You **do not** have to know why or how NMR spectra come about, but it would aid your understanding and interpretation of this type of spectroscopy if you knew a few background facts.

There are FOUR main features of an NMR spectrum:

1. the position of the absorbance peaks (the δ value)
2. the number of peaks
3. the relative peak area. This is shown by an integration value.
4. the splitting of each absorbance peak into smaller, finer peaks or multiplets. For example a peak split into two smaller peaks is called a doublet, one split into three peaks a triplet, etc.

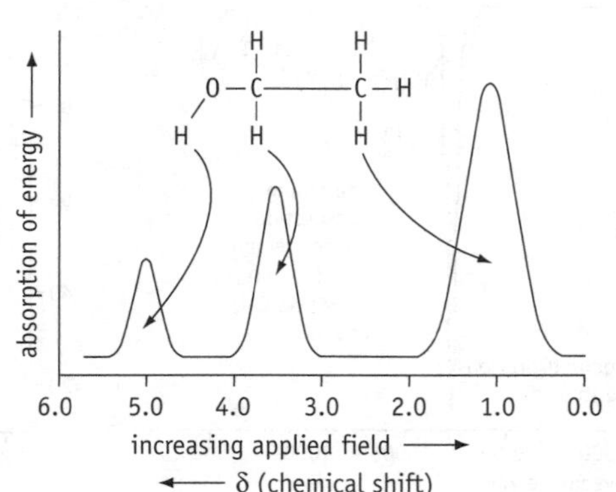

An example of a low-resolution (no splitting shown) spectrum, is that of ethanol, $\mathbf{CH_3CH_2OH}$ (see diagram). This spectrum has three peaks because of the three types of proton present and their positions in the spectrum are indicative of their position in the molecule relative to the oxygen of the –OH group. The areas under the three peaks also show the relative numbers of each type of proton i.e. 3 for the CH_3 protons, 2 for the CH_2 protons and 1 for the OH proton.

NMR is a very powerful technique. The table below summarises what it can tell you about a compound and how this is shown in its NMR spectrum.

What it tells you	How this is shown in the spectrum
How many types of proton present	The number of distinct peaks = number of types of proton
How many of each type of proton	The relative area under each peak = relative number of protons
What type of proton	The value of δ and then use of tables. See below
The number of chemically different protons adjacent to a particular type of proton	By the splitting of the distinct peaks. If there are *n* chemically different protons next to a particular type of proton then the peak for that proton is split into ***n* + 1** peaks

✓ Quick check 1

How can we relate NMR spectra to structures?

Example: Here are the spectra of two compounds having the molecular formula C_2H_6O.

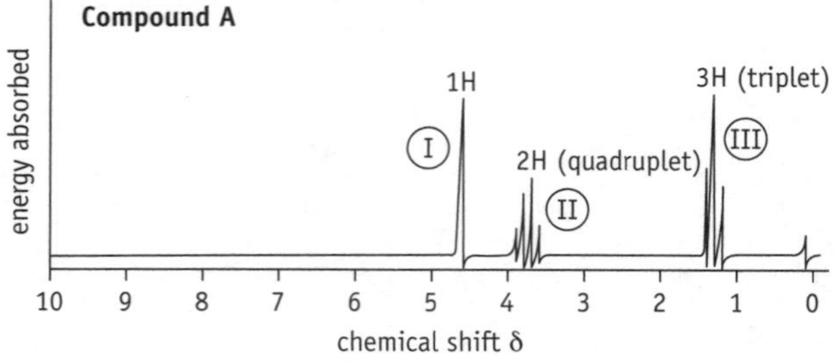

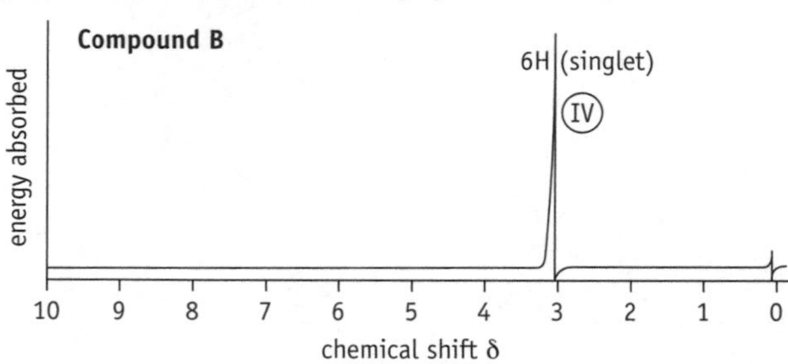

The steps we can take in explaining the spectra are shown below:

Step 1: Draw the displayed formulae of the two isomers, identifying the different types of proton present.

Step 2: Count the number of peaks. Compound A has three distinct sets of peaks and therefore three sets of chemically different protons. Therefore, compound A is ethanol. Similarly compound B has only one peak and therefore only one type of proton. Compound B is therefore methoxymethane.

ethanol

methoxymethane

Step 3: Explain the positions of the peaks. We can use our tables to identify these from their δ values. The protons for ethanol (compound A) are shifted differently because of their position in relation to the oxygen in the –OH group. The hydroxyl proton (labelled H_c) is therefore responsible for peak I (and there is only one of this type of proton δ = 4.5). Similarly protons H_b and H_c cause peaks II and III respectively. For compound B, proton type H_d gives rise to the absorbance of peak IV.

Table of δ values for different types of proton

Type of proton (in bold)	δ value
R-**CH**$_3$	0.7–1.6
R-**CH**$_2$-R	1.2–1.4
R$_3$**CH**	1.6–2.0
CH$_3$CO-R -CO-**CH**$_2$-R	2.1
C$_6$H$_5$-**CH**$_3$ C$_6$H$_5$-**CH**$_2$-CH$_3$	2.3–2.7
R-**CH**$_2$-Hal	3.2–3.7
R-O-**CH**$_3$ -O-**CH**$_2$-R	3.3–4.3
R-**CH**=CH$_2$	5.9
C$_6$H$_5$-O**H**	6.5–7.0*
C$_6$H$_5$-**H**	7.1–7.7
R-CHO	9.5–10*
-CO-OH	11.0–11.7*

NOTE these values vary and may not be the same for all compounds.
*These values are sensitive to solvent, substituents and concentration.

Step 4: Explain the splitting of the peaks.

Peak I – no splitting.

REASON – it hydrogen bonds to other molecules and only in extremely pure ethanol will you see splitting.

Peak II is split into 4 peaks.

REASON – it has THREE adjacent chemically different protons. Here n = 3 and therefore the peak is split into 4. The –O**H** proton will not cause splitting unless the ethanol is 100% pure (see above).

Peak III is split into THREE smaller peaks.

REASON – it has TWO adjacent chemically different protons. Here n = 2 and therefore the peak is split into 3.

Peak IV – there is no splitting.

REASON – there is only one type of proton.

✓ Quick check 2

Quick check questions

1 What information do the following features of an NMR spectrum tell you:
   (a) There are four distinct peaks.
   (b) One peak is split into a triplet.

2 The NMR spectrum shown is that for one of the isomers of the compound $C_2H_4Cl_2$.
   The number above the peak reflects its relative area.
   (a) Draw displayed formulae for the two possible isomers of the compound.
   (b) Name the isomer responsible for the spectrum.
   (c) Explain your answer to part (b).

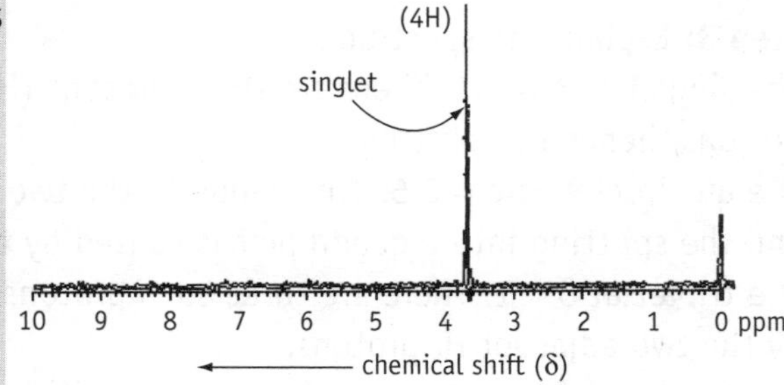

Identifying compounds using NMR spectroscopy

This section will show how powerful this method is in determining chemical structures.

Worked example

The NMR spectrum for an isomer of the compound C_4H_8O is given below. Identify the isomer and explain your reasoning.

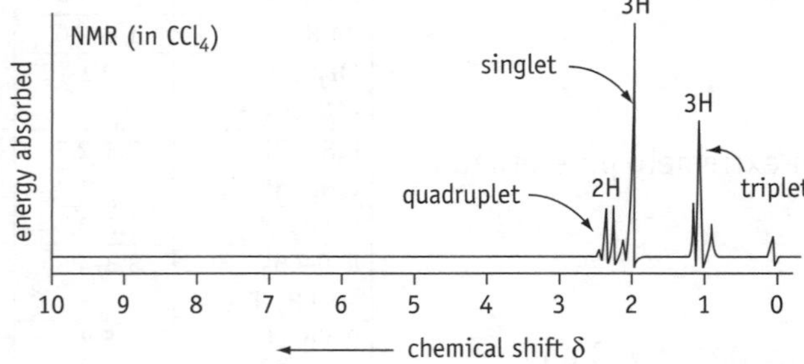

Step 1: Draw the displayed formulae for all the possible isomers of the compound.

butan-2-one

butanal

2-methyl propanal

Step 2: Identify the compound.
The three peaks in the spectrum show that there are three types of proton. Butan-2-one has three types (H_a, H_b and H_c) and 2-methylpropanal also has three (types H_h,H_i and H_j). Butanal, however, has four types (H_d, H_e, H_f and H_g). This narrows the choice to butan-2-one and 2-methylpropanal. However, none of the peaks indicate the presence of a proton corresponding to an aldehyde proton (δ = 9.5–10) and therefore the compound must be **butan-2-one.**

Step 3: Explain the spectrum.
The singlet at $\delta = 2.1$. There are three protons (H_a) and there are no adjacent protons, hence no splitting.
The quadruplet at $\delta = 2.5$. This is due to the two (relative peak area = 2) H_e protons and the splitting into a quadruplet is caused by the three adjacent protons, H_b.
The triplet at $\delta = 1$. There are three such protons and the peak is split into a triplet by the two adjacent H_b protons.
The δ values of the protons reflect their position with respect to the carbonyl oxygen.

✓ Quick check 1

Labile protons

- Protons attached to –OH, –COOH and $-NH_2$ groups can interchange with the protons in water. They are called **labile protons.**

- Adding the compound in question to deuterium oxide, D_2O, can identify such protons. For example, with an alcohol, ROH, the following exchange occurs.

ROH	$+D_2O \rightleftharpoons$	ROD	+HOD
peak present at $\delta = 4.5$		no peak at $\delta = 4.5$	

- The proton responsible for one of the peaks in the NMR spectrum of the alcohol has been removed and the peak itself disappears.
- This will not happen with protons that do not exchange with water such as methyl, $-CH_3$, protons.

> Deuterium (^2H) is an isotope of hydrogen and does not absorb at the same frequency range as the usual protons.

Quick check questions

Note: For all the spectra given below, the peak at $\delta = 0$ is due to the absorbance of the standard protons on TMS.

> TMS is the abbreviation for tetramethylsilane, a common solvent used in NMR measurements.

1 There are two possible isomers of the compound C_3H_6O.

(a) Draw displayed formulae for these two isomers.

(b) The NMR spectrum for one of these isomers is shown below. The numbers above each peak show the splitting and reflect the relative peak areas. Name the isomer responsible for the spectrum and explain your answer.

> The numbers in brackets refer to the numbers of hydrogen atoms.

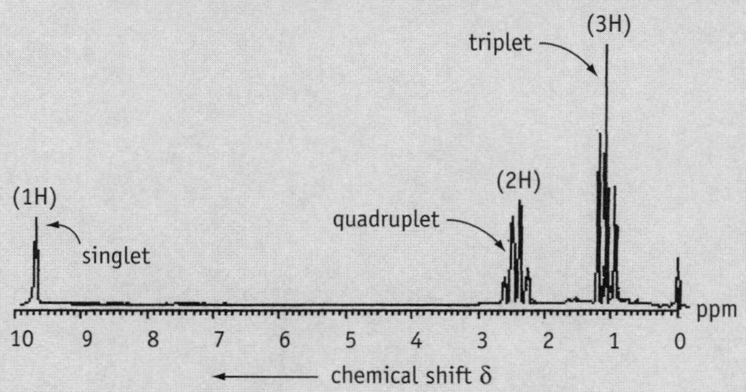

(c) Explain the number of peaks and the splitting patterns.

2 The NMR spectrum below is that of ethanoic acid.

(a) Draw the displayed formula of the compound.

(b) Explain why there are two peaks.

(c) Which peak is for which type of proton?

(d) Which peak will disappear if the acid is mixed with D_2O?

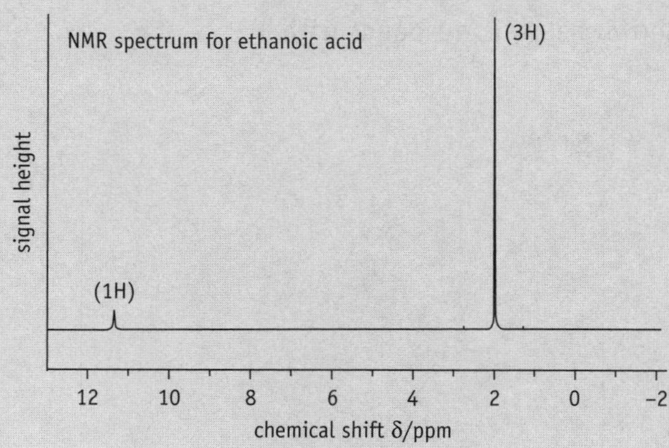

Mass spectroscopy of organic compounds

This is very straightforward because it is only used to find the relative molecular mass of the compound.

- We only have to look for one thing on the mass spectrum of any particular compound.
- This is the molecular ion peak (the M peak) for the molecules containing ^{12}C. This peak is the last but one of all the peaks on the mass spectrum.
- The last peak is a very small one caused by the presence of the M+1 ion This is due to the presence of carbon-13 (^{13}C) and is consequently smaller than the M peak.

✓ Quick check 1

The mass spectrum for ethanol (CH_3CH_2OH) is shown below.

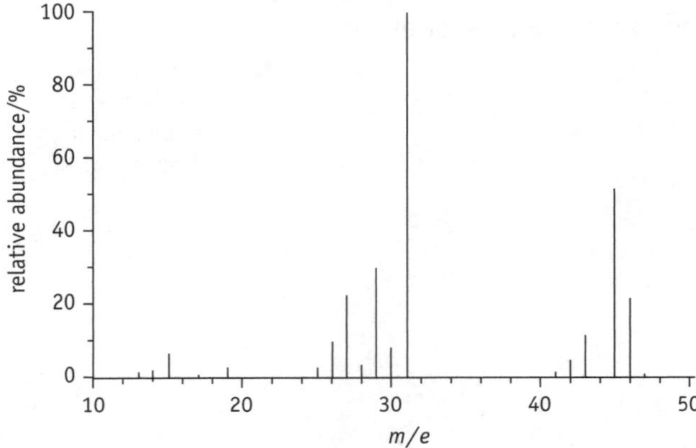

The **molecular ion peak** is the last but one peak at 46, and this corresponds to the relative molecular mass (2 x 12 + 6 x 1 +1 x 16).

✓ Quick check 2

Quick check questions

1 At what *m/e* value would you find the molecular ion peak for butan-2-one ($CH_3COCH_2CH_3$)?

2 Show that the mass spectrum of benzoic acid shown below corresponds with its structural formula of $C_6H_5CO_2H$.

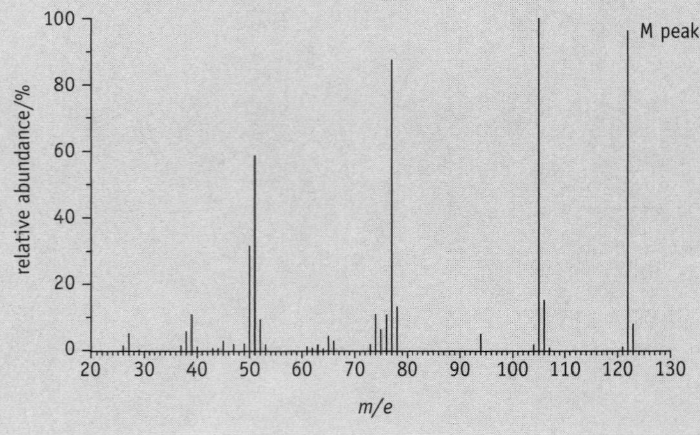

Unit D: End-of-unit exam questions

1 A hydrocarbon is known to contain a benzene ring. It has a relative molecular mass of 106 and has the following composition by mass: C, 90.56%; H, 9.44%.

a **i** Use the data above to show that the empirical formula is C_4H_5.

ii Deduce the molecular formula.

iii Draw structures for all possible isomers of this hydrocarbon that contain a benzene ring. [7]

b In the presence of a catalyst (such as aluminium chloride), one of these isomers A reacts with chlorine to give only **one** monochloro-product B.

i Deduce which of the isomers in (a)(iii) is A.

ii Draw the structure of B. [2]

2 The diagram below shows the three amino acids, glycine, alanine and phenylalanine.

```
                                    C6H5
                                     |
     H     O        CH3    O         CH2    O
     |    //         |    //          |    //
H2N—C—C        H2N—C—C         H2N—C—C
     |    \          |    \           |    \
     H     OH        H     OH         H     OH
  glycine          alanine         phenylalanine
```

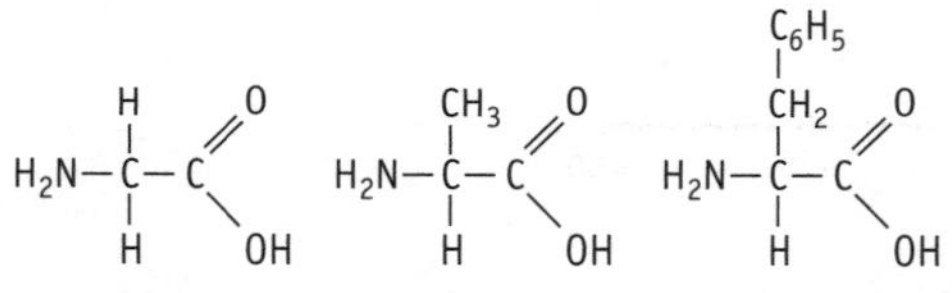

a Explain why glycine does **not** have optical isomers but the other two amino acids do. [1]

b Amino acids can react both with acids and with bases, and are also capable of forming zwitterions.

Write equations for the reactions between

i phenylalanine and HCl

ii alanine and NaOH. [2]

c The pH at which an amino acid forms its zwitterion is different for each amino acid and is known as the isoelectric point.

i The isoelectric point for glycine is at pH = 5.97. Draw the zwitterion formed by glycine at this pH. [1]

ii The isoelectric points of alanine and phenylalanine are at pH 6.00 and at pH 5.48 respectively. Draw the forms of the acids present at pH 5.7. [2]

d Glycine and alanine can react together to form the dipeptide shown below.

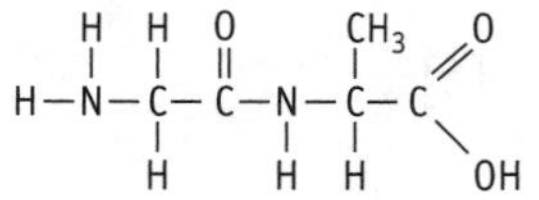

```
  H  H  O    CH3   O
  |  |  ||    |   //
H—N—C—C—N—C—C
     |     |  |   \
     H     H  H    OH
```

Draw the displayed formula of a different dipeptide that could be formed from the reaction between glycine and alanine. [1]

3 The medical drug amphetamine has the structure shown.

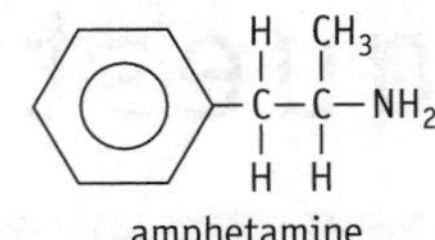

amphetamine

a Copy the diagram and use the symbol ":" to show where you would expect to find a lone pair of electrons in this structure. [1]

b Amphetamine is considerably more basic than phenylamine but is about as basic as ethylamine.

i Draw the displayed formulae of phenylamine and of ethylamine. [2]

ii Explain why you would expect amphetamine to be basic. [1]

iii Explain why phenylamine is a very weak base compared with ethylamine. [2]

iv Explain why amphetamine is more basic than phenylamine. [2]

4 The infrared and NMR spectra given below are for compound P, $C_4H_8O_2$.

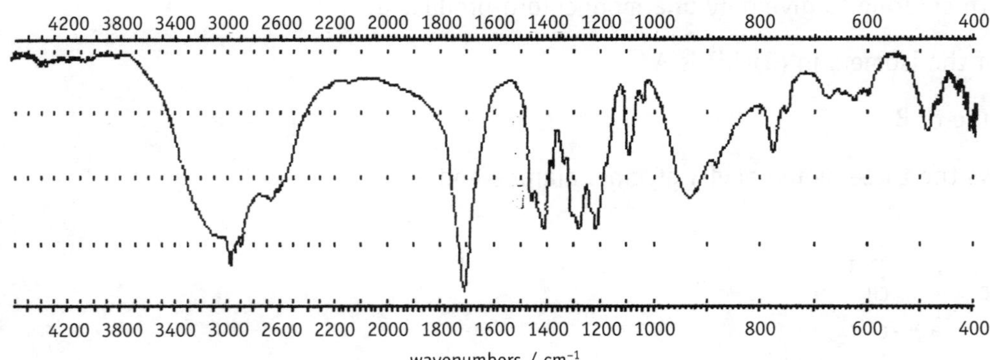

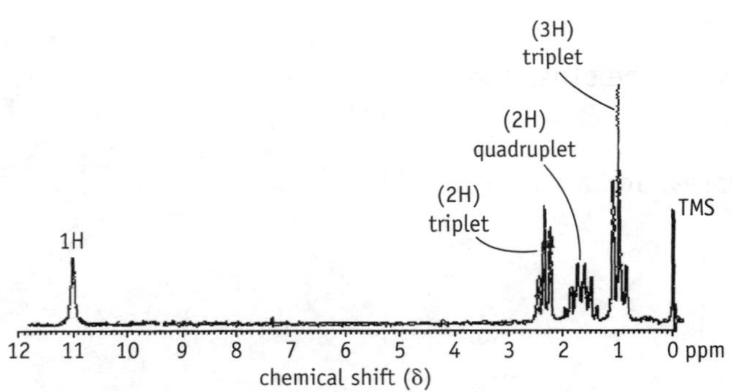

a Using the infrared spectrum deduce the functional group(s) present in P and **explain your answer**. [2]

b **i** How many types of proton are present in P?

ii Explain your answer. [2]

c Draw the five possible structural formulae for C_4H_8O and for each one state how many chemically distinct types of protons it possesses. [5]

d **i** Suggest which of the structures in part (c) is the correct structure for P.

ii Explain the NMR spectra for P, including the splitting patterns for each peak. [6]

Unit E: Trends and Patterns

This half module concentrates on inorganic chemistry. One important aspect of this examination is that it carries synoptic marks – it tests not just the content in the specification, but also links with other aspects of chemistry and previous knowledge you have acquired.

First of all you will look at a particular type of Hess cycle which involves the strength of an ionic lattice, and this will link with the work you have already done on structure and bonding. You will study the Periodic Table in more detail, and extend your understanding of inorganic compounds. In particular you will look at some compounds of the Period 3 elements, from sodium to argon. These compounds – the oxides and chlorides – show us how properties alter in the Periodic Table going across a period.

You will also investigate in some detail a whole new type of element – the transition elements – located in the d-block. You will look at the elements of the first row, from scandium to zinc, and their compounds. Transition element chemistry is very different from other inorganic element chemistry, so this section has a lot of new material in it, but also some familiar topics such as redox reactions and titrations.

Topic	**Reference to specification**	**Previous knowledge required**
Lattice enthalpy. Born–Haber cycles.	5.5.1	You will need to know thoroughly the work on enthalpy covered in the How Far, How Fast? module.
The Periodic Table: Period 3. Preparation of the oxides and chlorides of the Period 3 elements. Periodic trends of these elements.	5.5.2	In the Foundation module, there are 3 sections on the Periodic Table which you should know. These are the introduction, the Group 2 and Group 7 sections. It is also useful to revise the section on chemical bonding and structures.
Transition elements: • general properties • complex ions and ligands • precipitation reactions • redox chemistry • colorimetry • uv/visible spectroscopy	5.5.3	Most of this work is new, but the section in Foundation covering redox chemistry is essential background knowledge here.

Lattice enthalpy

You have already studied some enthalpy changes, ΔH, in How Far, How Fast? The lattice enthalpy is a particular type of enthalpy change for ionic solids only. It is the energy released when gaseous ions form a solid ionic lattice.

The lattice enthalpy, $\Delta H^{\ominus}_{latt}$ is the enthalpy change when 1 mole of an ionic compound is formed from its gaseous ions under standard conditions (298 K, 100 kPa).

$$M^+(g) + X^-(g) \rightarrow MX(s)$$

Don't forget the state symbols in this definition, especially the gaseous ions on the left.

Here is a checklist of things about the lattice enthalpy you must know. It is:

- always an *exothermic* change
- represented by the symbol $\Delta H^{\ominus}_{latt}$
- a measure of the strength of the ionic lattice in the compound – the ionic bond strength. A large exothermic value for the lattice enthalpy means a large electrostatic attraction between the ions in the lattice.
- $\Delta H^{\ominus}_{latt}$ cannot be measured by experiment, because gaseous ions do not combine directly to form a solid.

Trends in lattice enthalpy

The *size of the ions* and the *charge on the ions* affect the value of the lattice enthalpy. There are two trends in lattice enthalpy values that you need to know. The lattice enthalpy:

- is *more exothermic* if the *ionic charge increases*, because increased charge density on the ions means a stronger attraction between the ions and so a stronger lattice
- is *more exothermic* if the *ionic radius decreases*, for the same reason.

The more exothermic the enthalpy change is, the stronger the ionic bond. So if we compare the lattice enthalpies of sodium chloride and sodium iodide, we can see which has the strongest electrostatic attraction between ions:

$$\Delta H^{\ominus}_{latt}\text{(NaCl)} = -781\ \text{kJ mol}^{-1} \qquad \Delta H^{\ominus}_{latt}\text{(NaI)} = -699\ \text{kJ mol}^{-1}$$

NaCl has the more exothermic lattice enthalpy so has the stronger ionic bonding.

✓ *Quick check 1*

How to write lattice enthalpy equations

You need to be able to write an equation for the reaction which describes a lattice enthalpy. Look again at the definition of lattice enthalpy – when you write an equation representing a lattice enthalpy, remember that only *one mole* of the solid is formed. Balance the rest of the equation to fit in with this.

State symbols are important here, as gaseous ions react to form a solid. This reaction cannot be done in real life, but we still like to know the value of the lattice enthalpy because it gives useful information about the strength of the lattice.

So, the equation representing the lattice enthalpy of sodium chloride is:

$$Na^+(g) + Cl^-(g) \rightarrow NaCl(s)$$

and for the lattice enthalpy of magnesium oxide the equation is:

$$Mg^{2+}(g) + O^{2-}(g) \rightarrow MgO(s)$$

Worked example

Write the equation for the reaction showing the lattice enthalpy of sodium oxide.

Step 1: work out the formula of sodium oxide:

$$Na_2O$$

Step 2: see what the gaseous ions must be:

$$Na^+ \text{ and } O^{2-}$$

Step 3: put down these gaseous ions forming one mole of the compound:

$$Na^+(g) + O^{2-}(g) \rightarrow Na_2O(s)$$

Step 4: balance the equation. (REMEMBER there must be only 1 mole of $Na_2O(s)$):

$$2Na^+(g) + O^{2-}(g) \rightarrow Na_2O(s)$$

Step 5: check the state symbols are correct:

gaseous ions to solid compound

If you need reminding what the charges on different ions are, remember there is a list of these on page 3 of the AS revision book

Magnesium oxide, MgO, is used to line furnaces and protect the brickwork from the great heat inside the furnace. It can do this because it has a large exothermic lattice enthalpy, so has a very strong lattice which cannot be broken down by the heat. Magnesium oxide is a **refractory lining**.

✓ Quick check 2

✓ Quick check 3

? *Quick check questions*

1. Predict and explain which compound has the more exothermic lattice enthalpy, NaCl or CsCl.
2. Write an equation for the reaction representing the lattice enthalpy of aluminium chloride, $AlCl_3$.
3. Explain why magnesium oxide is used as a component of the ceramic used to make furnace rods.

The Born–Haber cycle

Lattice enthalpies can be calculated from a special type of Hess cycle called a **Born–Haber cycle**. Born–Haber cycles all have the same format. The OCR way of setting them out is shown here. Only one enthalpy change occurs in each step of the cycle, and this is shown in bold.

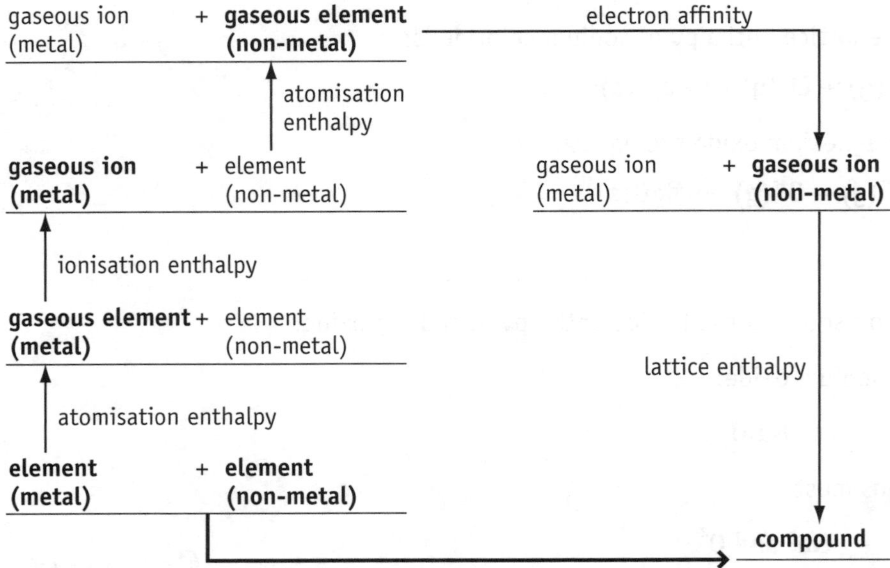

> This is a good place to revise general Hess cycle theory. Can you work out which two ways to go round the cycle?

As for all Hess cycles, there are two ways round.

- The first way round simply involves the enthalpy change of formation.
- The second way round involves *atomisation* (to get gaseous atoms) followed by *ionisation* (to get gaseous ions) and finally the *lattice enthalpy*.

So, according to Hess's law,

enthalpy change of formation = atomisation enthalpies + electron affinity + lattice enthalpy

and this can be rearranged to give

lattice enthalpy = enthalpy change of formation – atomisation enthalpies – electron affinity

You must be aware which energy changes are exothermic and which are endothermic, because exothermic changes go *down* in the Born–Haber cycle and endothermic changes go *up*.

exothermic	lattice enthalpy enthalpy change of formation electron affinity (for singly charged anions)
endothermic	enthalpy change of atomisation for both elements ionisation energy second electron affinity (for doubly charged anions)

The Born–Haber cycle for NaCl

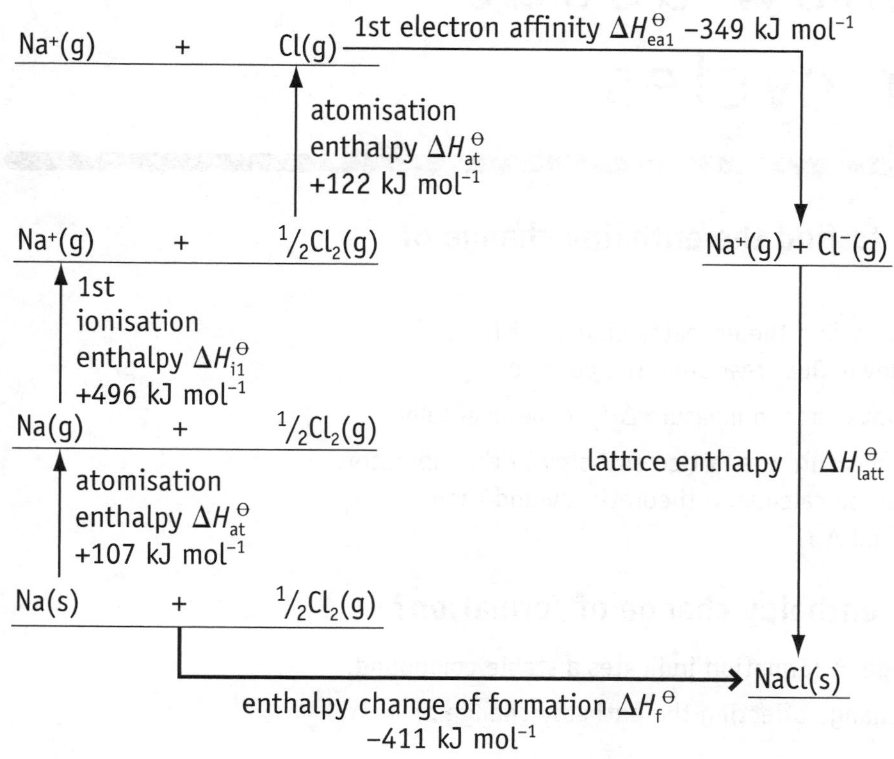

According to Hess's law,

lattice enthalpy $\Delta H^{\ominus}_{latt}$ = $\Delta H^{\ominus}_{f}$ (sodium chloride)
$-\Delta H^{\ominus}_{at}$ (sodium) – $\Delta H^{\ominus}_{i1}$ (sodium)
$-\Delta H^{\ominus}_{at}$ (chlorine) – $\Delta H^{\ominus}_{ea1}$ (chlorine)

Substituting,

lattice enthalpy = (–411) – (+107) – (+496) – (+122) – (–349) = **–787 kJ mol^{-1}**

Be careful here with the + and – signs before the numbers. It is best to use brackets as in this example.

Quick check question

1 Construct a labelled Born–Haber cycle for lithium fluoride.

Things to know about Born–Haber cycles

Using the Born–Haber cycle to find the enthalpy change of formation

- The Born–Haber cycle can be used to find the enthalpy change of formation, $\Delta H_f^\ominus$, if all the other terms are known. Just rearrange the equation.
- This can be useful, as it is not always easy to measure $\Delta H_f^\ominus$ experimentally.
- As you know, it is not possible to measure the lattice enthalpy in the laboratory either, but the lattice enthalpy can be calculated theoretically and then substituted into the equation to find $\Delta H_f^\ominus$.

What affects the size of the enthalpy change of formation?

- A large exothermic enthalpy change of formation indicates a stable compound.
- The largest exothermic enthalpy change affecting the enthalpy change of formation is the lattice enthalpy.
- The largest endothermic enthalpy change is the ionisation energy.

  As a rough guide, the balance between these two gives the final value for ΔH_f.

Get the ionisation energies right!

- Some compounds involve a metal ion with a 2+ charge, such as Mg^{2+}. The Born–Haber cycle in this case involves the *first ionisation enthalpy*, $\Delta H_{i1}^\ominus$ ($Mg(g) \rightarrow Mg^+(g) + e^-$) followed by the *second ionisation enthalpy*, $\Delta H_{i2}^\ominus$ ($Mg^+(g) \rightarrow Mg^{2+}(g) + e^-$). You can see this in the Born–Haber cycle for magnesium chloride opposite.
- Don't put $Mg(s) \rightarrow Mg^{2+}(g) + 2e^-$!!! Ionisation enthalpies involve removing *one mole of electrons at a time*.
- Remember that ionisation enthalpies have a positive value for ΔH. They are endothermic.

This is a good place to revise the definitions of different ionisation energies.

Are electron affinities endothermic or exothermic?

- Some compounds involve a non-metal ion with a 2– charge, such as O^{2-}.
- The Born–Haber cycle in this case involves:
  the *first electron affinity* $O(g) + e^- \rightarrow O^-(g)$ followed by
  the *second electron affinity* $O^-(g) + e^- \rightarrow O^{2-}(g)$.
- The first electron affinity is *exothermic* – energy is released when an electron is added.
- The second electron affinity is *endothermic* – energy must be added to get the negatively charged electron to add to the negatively charged ion.

The Born–Haber cycle for $MgCl_2$

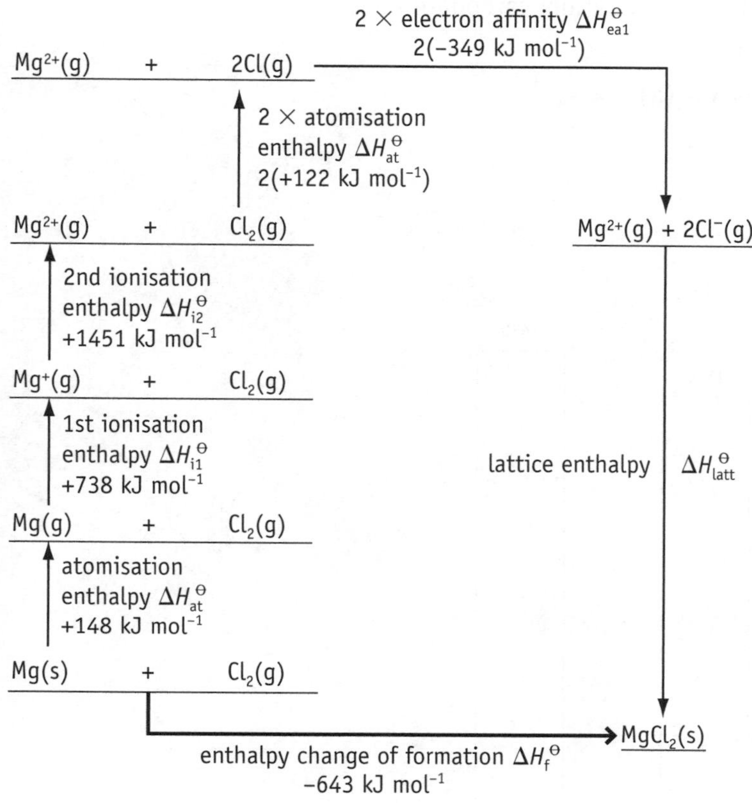

$\Delta H^{\ominus}_{latt}$ (magnesium chloride) = $\Delta H^{\ominus}_{f}$ (magnesium chloride)
$-\Delta H^{\ominus}_{at}$ (magnesium) − $\Delta H^{\ominus}_{i1}$(magnesium)
$-\Delta H^{\ominus}_{i2}$(magnesium) − $2 \times \Delta H^{\ominus}_{at}$ (chlorine)
$-2 \times \Delta H^{\ominus}_{ea1}$ (chlorine)

Substituting,

lattice enthalpy $\Delta H^{\ominus}_{latt}$ = (−643) − (+148) − (+738) − (+1451) − (+244) − (−698)
= −2526 kJ mol^{-1}

> Notice that the value for the atomisation enthalpy for chlorine is 2 × the value used in the NaCl Born–Haber cycle. This is because 2Cl(g) are needed in $MgCl_2$, instead of just Cl(g) in NaCl. The electron affinity is doubled too.

Quick check question

1 Calculate the enthalpy change of formation of calcium chloride using relevant enthalpy change values from the diagram above, plus:
atomisation enthalpy of calcium = +178 kJ mol^{-1}
1st ionisation enthalpy of calcium = +590 kJ mol^{-1}
2nd ionisation enthalpy of calcium = +1145 kJ mol^{-1}
lattice enthalpy of calcium chloride = −2258 kJ mol^{-1}

The decomposition of the Group 2 carbonates

Magnesium carbonate $MgCO_3$, calcium carbonate $CaCO_3$ and barium carbonate $BaCO_3$, all decompose in the same way when they are heated.

Example: $MgCO_3(s) \rightarrow MgO(s) + CO_2(g)$

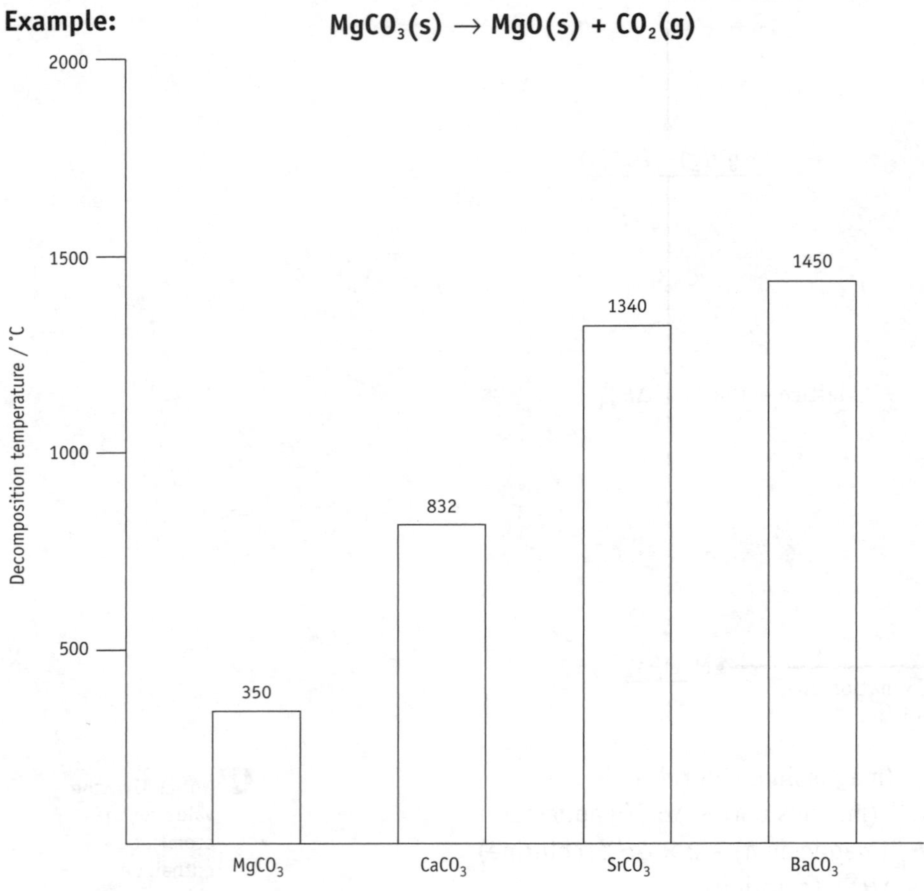

$MgCO_3$
$CaCO_3$
$BaCO_3$
decomposition temperature INCREASES

You can see that $MgCO_3$ has the lowest temperature of decomposition, and $BaCO_3$ has the highest.

This means **the trend is for increasing temperature of decomposition going down the Group.**

✓ Quick check 1

Think about the lattice enthalpies of these compounds. The magnesium ion has the smallest size, so $\Delta H^{\ominus}_{latt}$ (magnesium carbonate) is the most exothermic. This means (see p. 38) that $MgCO_3$ has the strongest lattice, and $BaCO_3$ the weakest lattice. So why does $MgCO_3$ decompose at the lowest temperature?

✓ Quick check 2

There are two reasons why the decomposition temperatures of the Group 2 carbonates increase going down the Group.

The main reason is the size of the cation.

- Mg^{2+} is the smallest cation, so has the highest charge density.
- This means that Mg^{2+} *distorts* or *polarises* the carbonate anion more than Ca^{2+} and Ba^{2+}. It pulls electrons onto one of the O atoms in the CO_3^{2-} anion – see the diagrams opposite. MgO and CO_2 are formed.
- It is easy for this to happen because the CO_3^{2-} anion is large with a *diffuse electron cloud*. It is *easily polarisable*.
- The structure of the lattice is weakened by this distortion. This means it breaks up more easily when heated.

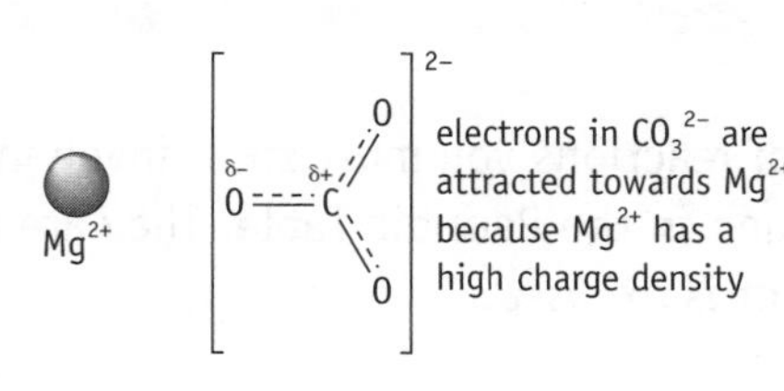

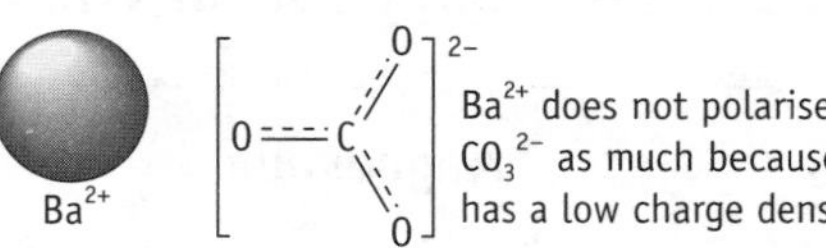

These carbonates have ionic bonding. The ions are held together by electrostatic charges in a giant lattice structure. This is a good place to revise the differences between ionic, covalent and metallic bonding – and the effect these structures have on properties.

Quick check 3

 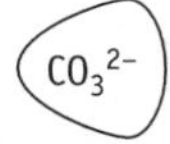 small Mg^{2+} ion polarises larger CO_3^{2-} ion

 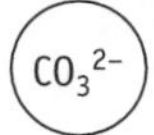 ions same size – no polarisation

The other reason is the type of oxide formed as the product.

- When the carbonates decompose, the corresponding oxides MgO, CaO and BaO are formed.
- The lattice enthalpy of MgO is the most exothermic, because the Mg^{2+} ion is the smallest.
- This means that MgO has a stronger lattice structure than CaO and BaO.
- The result is that MgO is a more favoured product than CaO and BaO so the formation of MgO is favoured.

Quick check 4

Quick check questions

1. State the trend in decomposition temperatures of the Group 2 carbonates. Give an equation showing this decomposition reaction.
2. State and explain the trend in lattice enthalpies of the Group 2 carbonates.
3. Explain why distortion of the lattice is greater in magnesium carbonate than in barium carbonate.
4. Explain, in terms of lattice enthalpies, why magnesium oxide has a stronger lattice than calcium oxide.

Reactions of some Period 3 elements with water and oxygen

There are 3 main groups of reactions you must know involving some of the elements of Period 3, sodium to argon in the Periodic Table. They are *redox reactions*, in which the Period 3 element is *oxidised*.

The reactions of sodium and magnesium with water

	Sodium	**Magnesium**
equation	$2Na(s) + 2H_2O(l) \rightarrow 2NaOH(aq) + H_2(g)$	$Mg(s) + 2H_2O(l) \rightarrow Mg(OH)_2(aq) + H_2(g)$
what is the reaction like?	• sodium fizzes about on the surface of the water and may catch fire, burning with an orange flame	• very slow with cold water but fast with steam • magnesium may appear not to react with cold water, but if it is left for a week, hydrogen can be collected and the solution is alkaline. • magnesium reacts rapidly with steam to give the hydroxide, but the main reaction here is • $Mg(s) + H_2O(g) \rightarrow MgO(s) + H_2(g)$
product	• sodium hydroxide is a strong alkali, pH = 12–14	• magnesium hydroxide is a weak alkali, pH 9–11

What are the trends here?

1 General reactivity

- Sodium reacts more vigorously with water than magnesium.
- This indicates a general trend – Group 1 elements are more reactive than Group 2 elements.

2 Alkali strength

- Sodium hydroxide is a stronger alkali than magnesium hydroxide.
- This is because sodium hydroxide is *more soluble* than magnesium hydroxide, so more OH^- ions are found in solution.

✓ Quick check 1

The reactions of magnesium, aluminium and sulphur with oxygen

	Magnesium	Aluminium	Sulphur
equation	$Mg(s) + \frac{1}{2}O_2(g) \rightarrow MgO(s)$ white magnesium oxide	$4Al(s) + 3O_2(g) \rightarrow 2Al_2O_3(s)$ white aluminium oxide	$S(s) + O_2(g) \rightarrow SO_2(g)$ sulphur dioxide **gas** $2S(s) + 3O_2(g) \rightleftharpoons 2SO_3(l)$ sulphur trioxide **liquid** **needs a heated platinum catalyst**
what is the reaction like?	magnesium burns vigorously once started with a bright white flame (used in fireworks and flares)	the reaction happens very easily at room temperature, and the layer of aluminium oxide seals the rest of the metal from attack by oxygen or water aluminium only catches fire and burns readily if it is powdered	sulphur burns easily with a blue flame to give sulphur dioxide gas, which can react with more oxygen to give sulphur trioxide gas
structure and bonding of the product	giant ionic lattice with ionic bonding	giant ionic lattice with ionic bonding	both SO_2 and SO_3 have simple molecules with covalent bonding
reaction of product with water	$MgO(s) + H_2O(l) \rightarrow Mg(OH)_2(aq)$ weak alkali pH 9–11 $Mg(OH)_2$ is a weak alkali because it partially ionises (dissociates) $Mg(OH)_2(aq) \rightleftharpoons Mg^{2+}(aq) + 2OH^-(aq)$	does not react with water	$SO_3(l) + H_2O(l) \rightarrow H_2SO_4(aq)$ sulphuric acid strong acid pH 1–2 this reaction is big business H_2SO_4 is a strong acid because it is fully dissociated $H_2SO_4(aq) \rightarrow 2H^+(aq) + SO_4^{2-}(aq)$

What are the trends here?

1 Bonding in the oxides

- Metals are found on the left of the Period, so the bonding in the oxides is *ionic*.
- The oxides of the non-metals on the right have *covalent bonding*.

ionic – – – – PERIOD 3 – – – – covalent

2 Structure of the oxides

- Metal oxides on the left of the Periodic Table have *giant lattice* structures.
- Non-metal oxides on the right of the Periodic Table have *simple molecular* structures.

giant lattice– – – – PERIOD 3 – – – –simple molecular

3 Acid/base nature of the oxides

- Metals on the left of the Period react with oxygen to give *basic oxides* which dissolve in water to give *alkaline solutions.*
- Non-metals on the right of the Period react with oxygen to give *acidic oxides* which dissolve in water to give *acidic solutions*.

basic oxides – – – – PERIOD 3 – – – – acidic oxides

✓ Quick check 2,3&4

Quick check questions

1 Write equations, including state symbols, to show how magnesium reacts with (a) cold water (b) steam.

2 Write an equation to show how aluminium reacts with oxygen. Describe the appearance of the product and its reaction with water.

3 Describe, using examples where appropriate, the trend in structure and bonding of the oxides of Period 3 elements.

4 Predict the pH of the solution formed when phosphorus(VI) oxide, PO_3, dissolves in water. Explain your answer.

Reactions of some Period 3 elements with chlorine

The reactions of sodium, magnesium, silicon and phosphorus with chlorine

	Sodium	**Magnesium**	**Silicon**	**Phosphorus**
equation	$2Na(s) + Cl_2(g) \rightarrow 2NaCl(s)$ white sodium chloride	$Mg(s) + Cl_2(g) \rightarrow MgCl_2(s)$ white magnesium chloride	$Si(s) + 2Cl_2(g) \rightarrow SiCl_4(l)$ volatile colourless liquid	$P_4(s) + 6Cl_2(g) \rightarrow 4PCl_3(l)$ volatile colourless liquid *then:* $PCl_3(l) + Cl_2(g) \rightarrow PCl_5(s)$ pale yellow solid
what is the reaction like?	• sodium burns in chlorine gas to give this reaction	• magnesium burns in chlorine gas to give this reaction	• silicon is heated in chlorine gas to give this reaction	• phosphorus combines directly with chlorine – it does not need heating • in excess chlorine only PCl_5 is formed
structure and bonding of the product	• giant ionic lattice with ionic bonding	• giant ionic lattice with ionic bonding	• simple molecules with covalent bonding	• liquid PCl_3 and vaporised PCl_5: simple molecules with covalent bonding • solid PCl_5: giant ionic lattice with ionic bonding between PCl_4^+ and PCl_6^- ions
reaction of the product with water	• dissolves to give a neutral solution, pH 7, containing $Na^+(aq)$ and $Cl^-(aq)$ ions	• dissolves to give a slightly acidic solution, pH 6.5, containing $Mg^{2+}(aq)$ and $Cl^-(aq)$ ions	• an exothermic, vigorous reaction **(hydrolysis)** giving off white fumes of hydrogen chloride, a white solid and an acidic solution, pH 2 $SiCl_4(l) + 4H_2O(l) \rightarrow Si(OH)_2(s) + 4HCl(g)$	• an exothermic, vigorous reaction **(hydrolysis)** giving off white fumes of hydrogen chloride and an acidic solution, pH 2 $PCl_3(l) + 3H_2O(l) \rightarrow H_3PO_3(aq) + 3HCl(g)$ phosphonic acid

What are the trends here?

1 Bonding in the chlorides

- Metals are found on the left of the Period, so the bonding in the chlorides is *ionic*.
- The chlorides of the non-metals on the right have *covalent bonding*.
- This is just the same as the bonding in the oxides!

ionic – – – – PERIOD 3 – – – – covalent

2 Structure of the chlorides

- Metal chlorides on the left of the Periodic Table have *giant lattice* structures.
- Non-metal chlorides on the right of the Periodic Table have *simple molecular* structures.
- This is just the same as the structures of the oxides!

giant lattice – – – – PERIOD 3 – – – –simple molecular

3 Acid/base nature of the chlorides

- The metal chlorides on the left of the Period dissolve to give *neutral* (or very slightly acidic) solutions.
- The non-metal chlorides on the right of the Period dissolve to give *acidic* solutions.

neutral chlorides– – – – PERIOD 3 – – – – acidic chlorides

? *Quick check questions*

1. State the structure and bonding in PCl_3 and PCl_5.
2. Describe the reaction of silicon tetrachloride with water.
3. Write an equation to show how magnesium forms magnesium chloride. Give the structure and bonding of the product of this reaction, and describe its reaction with water.
4. Give an equation for the hydrolysis of phosphorus trichloride. What is the final pH of this reaction?

Transition elements – electronic configurations

Transition elements are metals. You will only be concerned with the first row of transition elements, from titanium to copper. They are located in the d-block of the Periodic Table because the outermost electrons are in the d sub-shell.

There are many different things to learn about transition elements. Here is a checklist. Make sure you cover them all.

- definition of a transition element
- electronic configurations of the elements and their ions
- oxidation states
- catalytic behaviour
- hydroxides
- colour
- complex ions
- working out the formula of a complex using colorimetry
- redox reactions and titration calculations

Definition of a transition element

A transition element has at least one ion with a partly filled d orbital.

Learn this definition of a transition element.

Quick check 1

Electronic configuration of the transition elements

scandium	Sc	[Ar] $3d^1\ 4s^2$
titanium	Ti	[Ar] $3d^2\ 4s^2$
vanadium	V	[Ar] $3d^3\ 4s^2$
chromium	Cr	[Ar] $3d^5\ 4s^1$
manganese	Mn	[Ar] $3d^5\ 4s^2$
iron	Fe	[Ar] $3d^6\ 4s^2$
cobalt	Co	[Ar] $3d^7\ 4s^2$
nickel	Ni	[Ar] $3d^8\ 4s^2$
copper	Cu	[Ar] $3d^{10}\ 4s^1$
zinc	Zn	[Ar] $3d^{10}\ 4s^2$

Remember that *partly filled d orbital* means d^1 to d^9.

[Ar] is $1s^2\ 2s^2\ 2p^6\ 3s^2\ 3p^6$

Quick check 2

All the elements here have outermost electrons in the d sub-shell. The 3d sub-shell is filled after the 4s sub-shell, so most of the elements have a full 4s sub-shell, $4s^2$. *But note that chromium and copper have $4s^1$ electronic configurations, not $4s^2$*. This is to allow either a half-filled or a filled d sub-shell to be made – Cr has $3d^5\ 4s^1$ and Cu has $3d^{10}\ 4s^1$. A half filled or completely filled sub-shell is more stable, so it makes sense in energy terms for chromium and copper to have these electronic configurations.

> $3d^5$ and $3d^{10}$ are stable because the charge on the electrons is distributed symmetrically around the nucleus.

Zinc, and sometimes scandium, are not included in lists of transition metals although they are in the first row of the d-block. This is because:

- Zinc only forms one ion, Zn^{2+}, with an electronic configuration of $1s^2\ 2s^2\ 2p^6\ 3s^2\ 3p^6\ 3d^{10}$. This means that its ion has a full, not a partially full, d sub-shell – so zinc is not a transition element.
- Scandium forms one ion, Sc^{3+}, with an electronic configuration of $1s^2\ 2s^2\ 2p^6\ 3s^2\ 3p^6$. This is the main ion, and it has an empty d orbital, and so scandium is often excluded from the list of transition metals.
- We commonly say that scandium and zinc are not transition elements, but they are d-block elements. They are often included in the list of transition elements because their compounds are chemically similar in many (but not all!) ways to those of the other transition elements.

> If a d-block element has a white compound, in that oxidation state it will have a completely full or empty d orbital. *Colour* is associated with *partly filled d orbitals*.

Electronic configurations of the ions

The transition elements form positive ions as they are metals. This means that when an ion is formed, electrons are removed from the atom.

You must remember that the *4s electrons are removed first*.

Worked examples

1 Give the electronic configuration of Cu^{2+}.

Step 1: write down the electronic configuration of the atom, Cu.
[Ar] $3d^{10}\ 4s^1$

Step 2: The ion has a +2 charge so 2 electrons are removed. One is taken from the 4s sub-shell, the other from the 3d sub-shell.
So the electronic configuration of Cu^{2+} is: $1s^2\ 2s^2\ 2p^6\ 3s^2\ 3p^6\ 3d^9$

2 Give the electronic configuration of Fe^{3+}.

Step 1: write down the electronic configuration of the atom, Fe.
[Ar] $3d^6\ 4s^2$

Step 2: The ion has a +3 charge so 3 electrons are removed. Two are taken from the 4s sub-shell, the other from the 3d sub-shell.
So the electronic configuration of Fe^{3+} is: $1s^2\ 2s^2\ 2p^6\ 3s^2\ 3p^6\ 3d^5$

✓ Quick check 3

? *Quick check questions*

1. Explain what is meant by the term *transition element*.
2. Give the electronic configurations of manganese, Mn, and chromium, Cr.
3. Give the electronic configurations of Mn^{2+} and Cr^{3+}.

Transition elements – oxidation states, catalytic behaviour and the hydroxides

Oxidation states

One characteristic of the transition elements is that they have more than one oxidation state. The reason for this concerns the electrons in the d orbitals. The electrons in the d-sub-shell are close together in energy, so it is almost as easy to remove several outer electrons as it is to remove just one. In other elements there is a large energy difference between outer electrons in s or p orbitals, so usually only one type of ion is formed.

You have to learn the different oxidation states of iron and copper.

- **Iron** can have +2, +3, +4, +5 and +6 oxidation states. The +2 and +3 oxidation states are the most common, so the Fe^{2+} and Fe^{3+} ions are the most common.
- **Copper** can have +1, +2 and +3 oxidation states. The +1 and +2 oxidation states are the most common, so the Cu^{+} and Cu^{2+} ions are the most common.

Learn these common oxidation states:
for Fe, +2, +3
for Cu, +1, +2

Quick check 1

Catalytic behaviour

Transition metals and their compounds are very good catalysts. There are two reasons for this:

- They can have different oxidation states, which means they can gain and lose electrons easily. So they can transfer electrons to make a reaction go faster.
- They provide sites at which reactions can take place, because they bond to a wide range of ions and molecules in solution and as solids.

Examples of industrial catalysts are:

finely divided **Fe** or **Fe_2O_3** in the production of ammonia

$$N_2(g) + 3H_2(g) \rightleftharpoons 2NH_3(g)$$

solid **V_2O_5** in the production of sulphuric acid

$$2SO_2(g) + O_2(g) \rightleftharpoons 2SO_3(g)$$

finely divided **Ni** in the hydrogenation of alkenes

$$CH_2{=}CH_2(g) + H_2 \rightarrow CH_3CH_3(g)$$

Quick check 2

Transition metal hydroxides

Transition metal ions react with the hydroxide ion in aqueous solution to give a solid.

NaOH(aq) + transition metal ion(aq) → metal hydroxide(s)

> The NaOH(aq) supplies the OH^- ions

The colour of the metal hydroxide can be used to *identify the metal.*

You need only know about copper(II) hydroxide, iron(II) hydroxide and iron(III) hydroxide.

> Make sure you can write a full equation, as well as the ionic equations shown here, for these reactions.

$\mathbf{Cu^{2+}(aq) + 2OH^-(aq) \rightarrow Cu(OH)_2(s)}$	blue
$\mathbf{Fe^{2+}(aq) + 2OH^-(aq) \rightarrow Fe(OH)_2(s)}$	green
$\mathbf{Fe^{3+}(aq) + 3OH^-(aq) \rightarrow Fe(OH)_3(s)}$	rust

Example:

The reaction between aqueous copper sulphate and aqueous sodium hydroxide is

$$\mathbf{CuSO_4(aq) + 2NaOH(aq) \rightarrow Cu(OH)_2(s) + Na_2SO_4(aq)}$$

Although these reactions are the only ones you must know, make sure you can predict the reaction between any transition metal ion and aqueous sodium hydroxide.

Worked example

Predict the formula of the precipitate formed when $CoCl_3$ is dissolved in aqueous sodium hydroxide.

Step 1: Work out the formula of the transition metal ion:
In $CoCl_3$ the cobalt ion must be Co^{3+}.

Step 2: The charge on the metal ion is balanced by the number of hydroxide ions in the precipitate:
The charge is 3+ so three OH^- ions are needed.

Step 3: Write the ionic equation:
$\mathbf{Co^{3+}(aq) + 3OH^-(aq) \rightarrow Co(OH)_3(s)}$

✓ *Quick check 3*

Quick check questions

1 Discuss the variable oxidation states of transition metals.

2 Suggest why *finely divided* iron is used as a catalyst in the production of ammonia.

3 A transition metal ion in aqueous solution was added to aqueous sodium hydroxide. A green precipitate appeared. Identify the transition metal and write an ionic equation showing the reaction.

Transition elements and colour

Coloured ions

Transition metal ions are often coloured in aqueous solution. The reason for this colour concerns the electrons in the d orbitals – you will learn more about this if you take the Option called Transition Elements. For this part of the specification, it is enough for you to know that *colour is associated with a 3d sub-shell that is only partly filled.*

Learn these examples:

Ion	Colour in solution	Electronic configuration	Example
Fe^{2+}	green	[Ar] $3d^6$	$FeSO_4$
Fe^{3+}	yellow	[Ar] $3d^5$	$FeCl_3$
Cu^{2+}	blue	[Ar] $3d^9$	$CuSO_4$

In each case you can see that the 3d sub-shell is partly filled.

Now look at these more unusual cases:

- The Cu^+ ion in aqueous solution is colourless because its electronic configuration is $1s^2 2s^2 2p^6 3s^2 3p^6 3d^{10}$. It has a full 3d sub-shell so is not coloured like other transition metal ions.
- The ion Zn^{2+} also has the electronic configuration $1s^2 2s^2 2p^6 3s^2 3p^6 3d^{10}$, so is also colourless in solution.
- The ion Sc^{3+} has the electronic configuration $1s^2 2s^2 2p^6 3s^2 3p^6$, so has an empty 3d sub-shell . It is also colourless in solution.
- Solid compounds of Cu^+, Zn^{2+} and Sc^{3+} are white.

✓ Quick check 1

Colour

When visible light falls on a coloured object, certain wavelengths (or colours) are absorbed by the object.

We can only see the colours which remain – they are called the transmitted colours.

Visible light contains red, orange, yellow, green, blue and violet colours. If an object absorbs all wavelengths except green, the transmitted light is green and enters our eyes, so we see the colour green.

We use a colour wheel to help decide which colour an object is. If a substance absorbs one of these colours, the colour that is transmitted and that we see is the colour opposite it in the colour wheel. This is called the *complementary colour*.

So if an object absorbs red light only, it will appear green.

The colour wheel shows the primary colours red, blue and yellow plus the secondary colours violet, green and orange.

Ultra-violet/visible spectroscopy

If a substance is coloured, light in the visible region is involved. We can measure this light by a technique called **spectroscopy**, and often the ultra-violet (uv) as well as the visible region is taken. A **spectrometer** is the machine which measures the uv/visible light. It produces a recording called a **spectrum**.

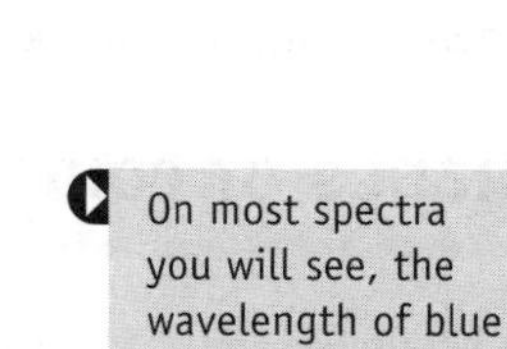

The diagram shows a typical visible spectrum of a transition metal ion in solution. The wavelength of the light is shown on the *x*-axis, with units of nanometres, nm. The *y*-axis shows the absorption, and there are usually no units here. If the line rises, then it means that light of that particular wavelength is absorbed.

You need to be able to tell from the spectrum what colour the transition metal ion is. So you must remember:

- wavelength of blue light is 450 nm
- wavelength of yellow light is 580 nm (just below 600 nm)
- wavelength of red light is 650 nm

On most spectra you will see, the wavelength of blue light is on the left, and red light is on the right.

You can see from the spectrum which wavelength is absorbed, and you will be able to work out which colour is absorbed. Then you can look up the colour opposite in the colour wheel to find out the colour you actually see.

Worked example

The spectrum above is that of aqueous Ti^{3+}. What colour is the solution?

Step 1: At what wavelength is the maximum absorption?
This spectrum shows strong absorption at 500–600 nm.

Step 2: What colour is light of this wavelength?
500–600 nm is yellow-orange light.

Step 3: Look at the colour wheel to see which colour is opposite yellow and orange. This is the colour you see.
Violet and blue are opposite yellow and orange. So the solution looks purple.

Quick check questions

1. Compounds of copper are usually coloured, yet CuCl is colourless when dissolved in water. Explain why this is so.
2. The diagram shows the visible spectrum of aqueous V^{3+}. State the colour of the solution and explain how you arrived at your answer.

✓ Quick check 2

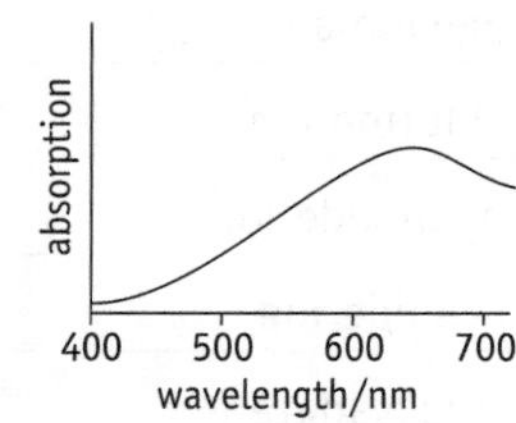

Complex ions

A **complex ion** is a *transition metal atom or ion + ligands.*

A **ligand** is a molecule or ion which *donates a pair of electrons* to the transition metal ion in a complex.

An example is $[Cu(H_2O)_6]^{2+}$.

This formula is used to show that Cu^{2+} is surrounded by six H_2O molecules. The H_2O molecules are the ligands. Square brackets go round the whole complex and the total charge of the complex ion goes outside these square brackets. The ligands are shown in normal brackets, with the number of ligands at the end.

Quick check 1

Shapes of complex ions

You will come across two different shapes of complex ions:

- octahedral complexes have six ligands
- tetrahedral complexes have four ligands

Quick check 2

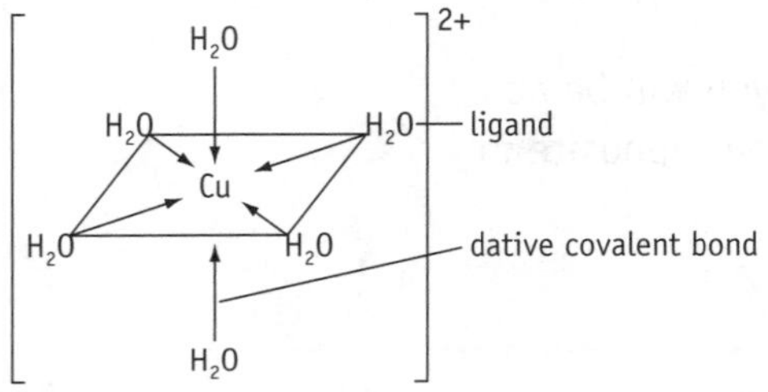

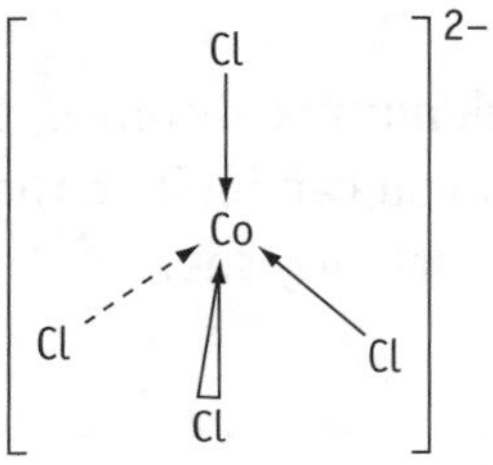

H_2O H_2O H_2O — ligand Cu H_2O H_2O — dative covalent bond H_2O $2+$

Cl Co Cl Cl Cl $2-$

When you draw a complex ion, make sure you show clearly which atom in the ligand forms the dative bond.

Types of ligands

- Ligands can be neutral or anionic (negatively charged).
- Ligands can donate **one** pair of electrons (**mono**dentate), **two** pairs of electrons (**bi**dentate) or **several** pairs of electrons (**poly**dentate).

Here are the most common ligands you will come across.

water	H_2O	monodentate
ammonia	NH_3	monodentate
chloride ion	Cl^-	monodentate
hydroxide ion	OH^-	monodentate
cyanide ion	CN^-	monodentate
thiocyanate ion	SCN^-	monodentate
ethane-1,2-diamine (en)	$H_2NCH_2CH_2NH_2$	bidentate
edta (ethylenediaminetetraacetic acid)		polydentate

edta is commonly used in shampoos and other cleaning agents

Ligand substitution

Some ligands combine more strongly with transition metal ions than others. A ligand that binds strongly can displace a ligand that binds more weakly. This is called **ligand substitution**. You can see ligand substitution in experiments because *different ligands change the colour of the metal ion* in solution.

Learn this useful trend in ligand strength:

edta > NH_3 > Cl^- > H_2O

There are certain ligand substitution reactions you must know:

- If excess aqueous ammonia is added dropwise to a solution of copper sulphate, the blue colour turns deep blue. This is because the copper sulphate solution contains the complex $[Cu(H_2O)_6]^{2+}$, which is the familiar blue colour (the sulphate ion is simply a spectator ion). As the NH_3 is added four H_2O ligands are replaced by four NH_3 ligands, and this new complex is a deeper blue colour.

$$[Cu(H_2O)_6]^{2+} + 4NH_3 \rightarrow [Cu(NH_3)_4(H_2O)_2]^{2+} + 4H_2O$$

blue **deep blue**

Both these complexes have six ligands, so they are octahedral in shape

- If excess hydrochloric acid is added dropwise to a solution of copper sulphate, the blue colour turns yellow. Here four Cl^- ligands replace all the H_2O ligands.

$$[Cu(H_2O)_6]^{2+} + 4Cl^- \rightarrow [Cu(Cl)_4]^{2-} + 6H_2O$$

blue **yellow**

Now, NH_3 is a stronger ligand than Cl^-, so it can substitute for the Cl^- in $[Cu(Cl)_4]^{2-}$

$$[Cu(Cl)_4]^{2-} + 4NH_3 + 2H_2O \rightarrow [Cu(NH_3)_4(H_2O)_2]^{2+} + 4Cl^-$$

yellow **deep blue**

✓ Quick check 3

- When aqueous potassium thiocyanate, KSCN, is added to a solution of Fe^{3+} ions, the complex formed is blood-red. One H_2O ligand is replaced by one SCN^- ligand.

$$[Fe(H_2O)_6]^{3+} + SCN^- \rightarrow [Fe(SCN)(H_2O)_5]^{2+} + H_2O$$

blood red

To get the overall charge on the complex ion you add together the charge on the metal ion and the charges on the ligands.
Overall charge = charge on metal + charges on all

? Quick check questions

1 Explain what is meant by the term "ligand". Give an example of a ligand.

2 Predict the shape of the following complex ions:
(a) $[Ni(NH_3)_6]^{2+}$; (b) $[Ni(en)_3]^{2+}$; (c) $[Ni(Br)_4]^{2-}$

3 Write equations to show what happens when concentrated hydrochloric acid is added to copper sulphate solution, followed by concentrated ammonia solution. Indicate any colour changes.

Colorimetry and the formulae of complex ions

Colorimetry is a technique which allows you to work out the formula of a complex ion. The equipment used is called a **colorimeter**.

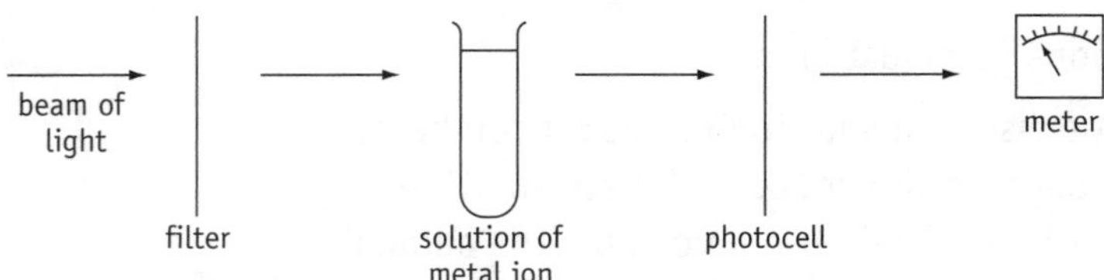

- In a colorimeter a narrow beam of light passes through the solution (containing the complex) and hits a *photocell* which records the wavelength of light transmitted – and hence the colour.
- You can choose which colour is measured by changing the *filter* the light passes through before it reaches the solution.
- You make up a solution of a metal ion, so that it forms a complex with water as the ligand. Then you gradually add a second ligand in solution which substitutes for the H_2O and gives a different colour. You will work out how many of these new ligands are present in the final complex.
- As the second ligand is added, the colour of the solution changes gradually. When the ligand substitution is complete, the colour does not change any more.
- The colorimeter measures the colour, so when the reading stays constant you know that no more ligand substitution is taking place.

It is important that you can describe the experimental details for this type of experiment.

- Take a set of test-tubes, 10 or 11 is the usual number.
- Number them so you don't get them mixed up.
- In test tube 1 put water only. This is the control you use to set the colorimeter to zero.
- Now take a solution of the metal ion, and a separate solution of the ligand. You must know the concentrations of these so that you can work out how many moles of each are in the test tube.
- Put a different amount of each solution in each test tube, using a burette for accuracy.
- Make sure the *final volume in each test tube is the same* (if necessary add water).
- Usually the number of moles of metal ion remains constant and the number of moles of ligand increases (see graph A opposite).
- Sometimes the number of moles of each changes, so there is a different proportion of metal ion to ligand in each test tube (see graph B opposite).

- Take the test tube containing the deepest colour solution and put it into the colorimeter. Adjust the needle so that it is at the maximum end of the scale. This gives you readings over the greatest part of the scale, which is good practice as it increases the accuracy of the readings.
- Now take readings for all the test tubes and plot the results on a graph.

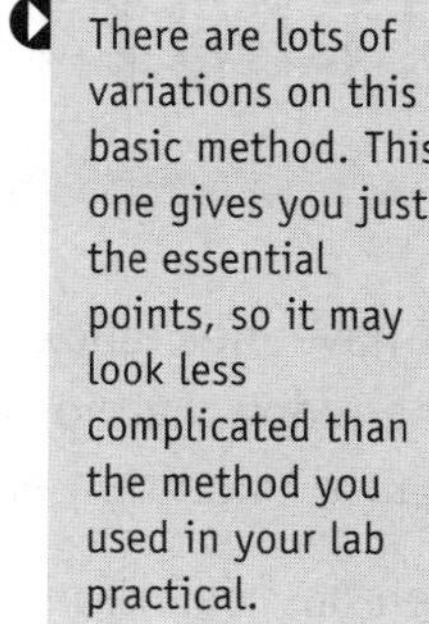

The graph has absorbance (no units) on the vertical axis, and test tube number on the horizontal axis.

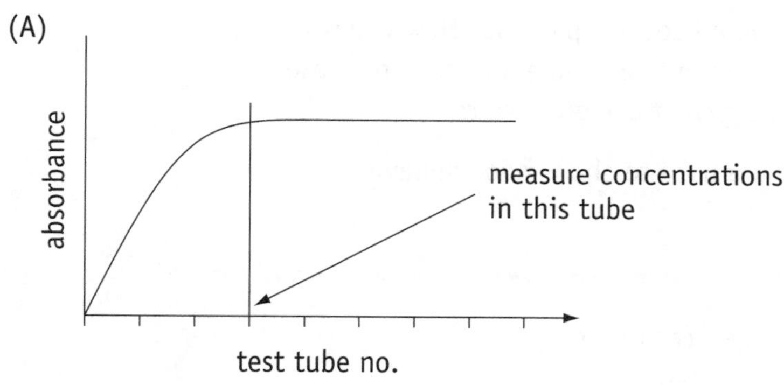

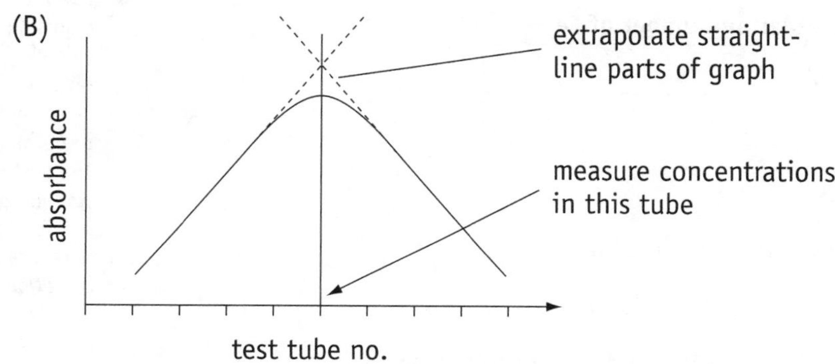

Work out the number of moles of metal ion and the number of moles of ligand in the test tubes indicated. This ratio gives you the formula of the ligand.

Quick check question

1 Work out the formula of the complex formed between Ni^{2+} and edta, using the following colorimeter readings:

test tube	1	2	3	4	5	6	7	8	9	10	11
cm^3 Ni^{2+}(aq) 0.1 mol dm^{-3}	0.0	0.5	1.0	1.5	2.0	2.5	3.0	3.5	4.0	4.5	5.0
cm^3 edta(aq) 0.1 mol dm^{-3}	5.0	4.5	4.0	3.5	3.0	2.5	2.0	1.5	1.0	0.5	0.0
absorbance	0.20	0.40	0.61	0.79	0.91	1.01	0.89	0.81	0.60	0.41	0.19

Redox reactions and titration calculations

You have met redox reactions before in the AS course (page 30) Now you will study some redox reactions involving transition metal ions. There are many of these, because transition elements have *several different oxidation states*.

Often *half equations* and *ionic equations* are used to show redox behaviour.

Redox behaviour in iron

$$\underset{\text{green}}{Fe^{2+}(aq)} \rightleftharpoons \underset{\text{yellow}}{Fe^{3+}(aq)} + e^-$$

species	oxidation number of Fe
Fe^{2+}	+2
Fe^{3+}	+3

Iron can change from

oxidation state +2 to +3 if an oxidising agent is added

oxidation state +3 to +2 if a reducing agent is added

- Fe^{3+} is itself an oxidising agent, because it can oxidise other species.
- Fe^{2+} is itself a reducing agent, because it can reduce other species.

Use the terms oxidising, reducing, oxidising agent and reducing agent carefully – it's easy to get muddled. The best way of working out what is oxidised and what is reduced is to use *oxidation numbers*.

Example:

$$Fe^{3+}(aq) + 2I^-(aq) \rightarrow Fe^{2+}(aq) + I_2(aq)$$

You can tell by the colour change that this reaction has happened. Fe^{3+} is yellow, Fe^{2+} is pale green and I_2 is orange-brown. The brownish colour of the I_2 masks the pale green of Fe^{2+}, so the colour change is yellow to orange-brown.

✓ Quick check 1

Redox behaviour in manganese

$$\underset{\substack{\text{manganate(VII) ion}\\ \text{purple}}}{MnO_4^-(aq)} + 8H^+(aq) + 5e^- \rightarrow \underset{\substack{\text{manganese(II) ion}\\ \text{very pale pink}}}{Mn^{2+}(aq)} + 4H_2O(l)$$

This reaction takes place in acid solution, as you can see by the presence of H^+.

species	oxidation number of Mn
MnO_4^-	+7
Mn^{2+}	+2

Manganese is reduced in this reaction, so the manganate(VII) ion is an oxidising agent.

Examples:

$$MnO_4^- + 5Fe^{2+} + 8H^+ \rightarrow Mn^{2+} + 5Fe^{3+} + 4H_2O$$

The colour change here is from purple (MnO_4^-) to yellow (Fe^{3+}) – the very pale pink of the Mn^{2+} does not show.

$$2MnO_4^- + 5C_2O_4^{2-} + 16H^+ \rightarrow 2Mn^{2+} + 10CO_2 + 8H_2O$$

ethanedioate ion

The oxidation number of carbon changes from +3 in $C_2O_4^{2-}$ to +4 in CO_2. The colour change here is from purple to colourless – the very pale pink of the Mn^{2+} does not show.

The state symbols have been left out of these equations to make them clearer.

Redox behaviour in chromium

Notice that all these reactions take place in acid solution.

$$Cr_2O_7^{2-}(aq) + 14H^+(aq) + 6e^- \rightarrow 2Cr^{3+}(aq) + 7H_2O(l)$$

dichromate(VI) ion — **orange**; **chromium(III) ion** — **green**

species	oxidation number of Cr
$Cr_2O_7^{2-}$	+6
Cr^{3+}	+3

This reaction is frequently used in organic chemistry, where the orange → green colour change tells you the organic substance has been oxidised (see page 9)

Titration calculations

See page 12 of Revise AS Chemistry for OCR Spec. A

You must revise *titration calculations* here because they will be tested!

- Make sure you are familiar with the redox titration between MnO_4^- and Fe^{2+} in aqueous acid solution.
- The equation for this reaction is:

$$MnO_4^-(aq) + 5Fe^{2+}(aq) + 8H^+(aq) \rightarrow Mn^{2+}(aq) + 5Fe^{3+}(aq) + 4H_2O(l)$$

- The *purple* aqueous MnO_4^- is added from the burette to the aqueous Fe^{2+}. It immediately goes *colourless* as it reacts with the Fe^{2+} (the very pale pink of Mn^{2+} and the pale yellow of dilute Fe^{3+} do not show)
- The end point is when all the MnO_4^- has reacted and a *permanent pink colour* can be seen.
- Remember the acid! It takes part in the reaction so without it the reaction will not happen!

Remember the colour change at the end-point is **colourless → pink**

✓ *Quick check 2*

Quick check questions

1 Hydrogen peroxide, H_2O_2, is an oxidising agent.

$H_2O_2(aq) + 2H^+(aq) + 2e^- \rightarrow 2H_2O(l)$

Construct the equation for the reaction between hydrogen peroxide and Fe^{2+} ions in aqueous solution.

2 A titration was performed in which 14.9 cm^3 of 0.02 mol dm^{-3} potassium manganate(VII) was needed to react with 100 cm^3 of aqueous acidified $FeCl_2$. Calculate the concentration of the $FeCl_2$ solution.

Unit E: End-of-unit exam questions

1 (a) A student prepared two chlorides of iron in the laboratory.

- In the first experiment, the student reacted with iron with an excess of hydrogen chloride gas forming a chloride A, with the composition by mass: Fe: 44.0%; Cl 56.0%.
- In the second experiment, the student formed 8.12 g of a chloride B by reacting 2.79 g of iron with an excess of chlorine.

(a) Identify compounds A and B. Include all your working and give equations for the two reactions. [A_r: Fe, 55.8; Cl, 35,5.] [4]

(b) The student dissolved compound B in water, to form an orange solution containing the complex ion C. Addition of aqueous thiocyanate ions, SCN^-(aq), to C gave a blood-red solution containing the complex ion D.

(i) Identify the ions C and D and give their formulae.

(ii) Explain the bonding present in C.

(iii) Explain how D is formed from C and explain the term "ligand exchange". [8]

(c) Chloride A has a much higher melting point (672°C) than chloride B (220°C). An aqueous solution of chloride A is neutral whereas that of B is acidic. Explain what this information suggests about the structure and bonding in A and B. [6]

2 The table below relates to the oxides of elements in Period 3 in the Periodic Table.

Oxide	Na_2O	MgO	Al_2O_3	SiO_2	P_4O_{10}	SO_3
Melting point °C	1275	2827	2017	1607	580	33
Bonding						
Structure						

(a) Complete the table using the following guidelines.

(i) Complete the **bonding** row using **only** the words**: ionic** or **covalent**.

(ii) Complete the **structure** row using only the words: **simple molecular** or **giant**.

(iii) Explain, in terms of forces, why MgO and SO_3 have widely differing melting points. [5]

(b) The oxides Na_2O and SO_3 were each added separately to water.

For each oxide, construct a balanced equation for its reaction with water and suggest the pH of the resulting solution.

(i) Na_2O (ii) SO_3 [4]

(c) The oxides MgO and SiO_2 were each melted and then tested for electrical conductivity. Compare and explain the conductivity of each molten oxide. [2]

Unit F: Unifying Concepts in Chemistry

This component builds strongly on the material covered in How Far, How Fast? It involves much more quantitative work, and contains more maths than any other unit. The exam paper has about 15 marks on extended writing, and 45 marks on structured questions. This is the last unit of your A-level course. Well done for getting so far and good luck!

Topic	**Reference to specification**	**Previous knowledge required**
How Fast? You will learn how the concentrations of the reactants are connected to the reaction rate using rate equations. You will also see how the rate equation shows us the rate-determining step of the reaction.	5.11.1	How Far, How Fast? 5.3.2 Reaction rates Most of the work you did on reaction rates involved looking at catalysts and how they operate, which is not extended here. But you also looked at how temperature affects reaction rate, and this is explained in more detail. You will have to understand and draw many graphs in this section.
How Far? In this section you will learn how to calculate equilibrium constants in terms of concentration and partial pressure. You will then examine how this equilibrium constant changes when the conditions of the equilibrium are changed.	5.11.2	How Far, How Fast? 5.3.3 Chemical Equilibria The most important thing here is to remember what an equilibrium is, and how it shifts according to Le Chatelier's principle.
Acids, bases and buffers Here you will learn how to calculate pH values for strong and weak acids, and strong bases. You will learn that acid equations are equilibria with special equilibrium constants. You will titrate strong and weak acids and bases against each other and discover that the titration graphs you can plot are different in each case. You will also take a closer look at indicators, and discover the chemical definition of a buffer solution.	5.11.3	How Far, How Fast? 5.3.3 Chemical Equilibria In How Far? How Fast? you studied acids as proton donors, and now you will combine that knowledge with all the work on equilibria that you have done.

How Fast?
The rate equation

This part of the unit deals with *rates of reaction*.

From previous work, you know that

$$\textbf{rate of reaction} = \frac{\textbf{change in concentration}}{\textbf{time}}$$

> Units of concentration = mol dm^{-3}
> units of rate = mol dm^{-3} s^{-1} or mol dm^{-3} min^{-1} etc.

The **rate equation** is a way of *calculating the rate of a chemical reaction.*

The rate equation has a general form for the reaction

A + B → products

$$\text{rate} = k[A]^m[B]^n$$

where

- A and B are the *reactants*
- [A] is the *initial* concentration of A
- [B] is the *initial* concentration of B
- ***k*** is a constant called **the rate constant. *k*** relates the rate of a chemical reaction with the reactant concentrations. *The larger the value of* k, *the faster the reaction goes.*
- *m* is the **order** of the reaction with respect to reactant A
- *n* is the **order** of the reaction with respect to reactant B

> Note that [A] and [B] are the concentrations at the ***beginning*** of the reaction. This means we are measuring the ***initial*** rate. Of course the rate of a reaction changes with time, getting slower towards the end of the reaction. Measuring the initial rate, the instant the reactants are mixed, means there is zero concentration of products which may cause a reverse reaction.

The order of a reactant is the power to which its concentration is raised in the rate equation.

The **total order** of the reaction is ***m* + *n***, the sum of the orders for each reactant.

Each reaction has its own rate equation. *This can only be worked out experimentally.*

Examples:

$SO_2Cl_2(g) \rightarrow SO_2(g) + Cl_2(g)$	$BrO_3^-(aq) + 5Br^-(aq) + 6H^+(aq) \rightarrow 3Br_2(aq) + 3H_2O(l)$
rate = $k[SO_2Cl_2(g)]$	**rate =** $k[BrO_3^-][Br^-][H^+]^2$
The rate is • first order with respect to SO_2Cl_2 • first order overall	The rate is • first order with respect to BrO_3^- • first order with respect to Br^- • second order with respect to H^+ • fourth order overall

✓ Quick check 1

Units of the rate constant

Because each reaction has its own rate equation, the units for k change according to the rate equation.

You can work out the units of k.

Step 1: Rearrange the rate equation to get k:

$$k = \frac{\text{rate}}{[\text{A}]^m[\text{B}]^n}$$

Step 2: Substitute the units into this equation:

Units for rate are mol dm^{-3} s^{-1}.

Units for [A] and [B] are mol dm^{-3}.

Step 3: Cancel any units that are the same on the top and bottom of the expression.

Worked examples

1 For the reaction

$(CH_3)_3CBr(l) + H_2O(l) \rightarrow (CH_3)_3COH(l) + HBr(aq)$

the rate equation is rate = $k[(CH_3)_3CBr]$.

This is a first-order rate equation. What are the units of k?

$$k = \frac{\text{rate}}{[(CH_3)_3CBr]} = \frac{\text{mol dm}^{-3}\text{s}^{-1}}{\text{mol dm}^{-3}} = \text{s}^{-1}$$

2 For the reaction $2ICl(g) + H_2(g) \rightarrow 2HCl(g) + I_2(g)$ the rate equation is rate = $k[ICl][H_2]$.

This is a second-order rate equation. What are the units of k?

$$k = \frac{\text{rate}}{[ICl][H_2]} = \frac{\text{mol dm}^{-3}\text{s}^{-1}}{\text{mol dm}^{-3} \times \text{mol dm}^{-3}}$$

$$= \frac{\text{s}^{-1}}{\text{mol dm}^{-3}} \text{ which is written as dm}^3 \text{ mol}^{-1} \text{ s}^{-1}.$$

You don't have to write $\frac{\text{s}^{-1}}{\text{mol dm}^{-3}}$ as dm^3 mol^{-1} s^{-1}. It's just a mathematical convention.

You can see that *the units of* k *depend on the order of the reaction*. You can work these out, or remember:

order	**rate equation**	**units of *k***
zero order	rate = $k[A]^0$	mol dm^{-3} s^{-1}
first order	rate = $k[A]^1$	s^{-1}
second order	rate = $k[A]^2$	dm^3 mol^{-1} s^{-1}

Quick check questions

1 For the rate equation rate = $k[NO][O_3]$,
(a) what is the order with respect to NO?
(b) what is the total order of the reaction?

2 What are the units of k for the reaction in question 1?

✓ Quick check 2

How Fast? Calculating *k*, the effect of *T* and reaction mechanisms

You must be able to calculate the rate of a reaction from the rate equation, given the value of *k* and initial concentrations. This simply means substituting into the rate equation.

As well as this, make sure you can calculate *k*, the rate constant, using the rate equation. This means you have to rearrange the rate equation to give: $k = \frac{\text{rate}}{[A]^m[B]^n}$

Worked example

For the reaction

$$2KI(aq) + K_2S_2O_8(aq) \rightarrow 2K_2SO_4(aq) + I_2(aq)$$

$$\text{rate} = k[KI][K_2S_2O_8].$$

The initial concentration of KI is 1.0×10^{-2} mol dm^{-3}, and the initial concentration of $K_2S_2O_8$ is 5.0×10^{-4} mol dm^{-3}.

The initial rate of disappearance of $K_2S_2O_8$ is 1.02×10^{-8} mol dm^{-3} s^{-1}.

Calculate the rate constant.

You can see that this type of question involves lots of information. Don't be put off – just sort it out carefully.

Step 1: work out the equation to calculate *k* by rearranging the rate equation:

$$\text{rate} = k[KI][K_2S_2O_8]$$

$$\text{so } k = \frac{\text{rate}}{[KI][K_2S_2O_8]}$$

Step 2: list the values you have been given for the rate, the concentration of KI and the concentration of $K_2S_2O_8$. Make sure the units are all OK.

rate = 1.02×10^{-8} mol dm^{-3} s^{-1}
[KI] = 1.0×10^{-2} mol dm^{-3}
[$K_2S_2O_8$] = 5.0×10^{-4} mol dm^{-3}

Step 3: substitute into equation for *k*:

$$k = \frac{1.02 \times 10^{-8}}{(1.0 \times 10^{-2})(5.0 \times 10^{-4})}$$

Step 4: calculate *k*:

$$k = 2.04 \times 10^{-4}$$

Step 5: work out the units for *k*:

$$\text{units are } \frac{\text{mol dm}^{-3}\ \text{s}^{-1}}{(\text{mol dm}^{-3})(\text{mol dm}^{-3})} \text{ which is dm}^3\ \text{mol}^{-1}\ \text{s}^{-1}$$

$$\text{final answer } k = 2.04 \times 10^{-4}\ \text{dm}^3\ \text{mol}^{-1}\ \text{s}^{-1}$$

The effect of a temperature change on *k*

When the temperature of a reaction is increased the molecules move faster, so there are more collisions and more chance of a product being formed.

- A rise in temperature *increases* the value of *k*. This means the reaction goes faster.
- A fall in temperature *decreases* the value of *k*. This means the reaction slows down.

large *k* = fast reaction, small *k* = slow reaction

Reaction mechanisms

A reaction mechanism is a description of a chemical reaction as a series of one-step processes.

One of these steps may occur much more slowly than the others. This step is called the **rate-determining step (rds)**, because it determines how fast the reaction goes.

How do we know what the rate-determining step is? Well, that particular reaction is shown by the rate law.

In the rate law the *order of reaction* with respect to a reactant indicates how many molecules of that reactant participate in the rate-determining step.
If a reactant is first order, one molecule reacts in the rds.
If a reactant is second order, two molecules react in the rds etc.

Look at this example to understand:

$$H_2\ (g) + 2ICl\ (g) \rightarrow I_2(g) + 2HCl(g)$$
$$\text{rate} = [H_2][ICl]$$

- The *stoichiometric equation* tells us that *overall*, 1 molecule of H_2 reacts with 2 molecules of ICl to give the products. But it does not tell us about the individual steps which make up the overall reaction. *There is no link between the stoichiometric equation and the reaction mechanism or the rate equation.*
- The rate equation tells us that the rate-determining step involves 1 molecule of H_2 and 1 molecule of ICl, because it is first order with respect to both.

The actual mechanism is:
first
$H_2 + ICl \rightarrow HI + HCl$
slow step
then
$HI + ICl \rightarrow I_2 + HCl$
fast step

In this case the rds is the first step, and an intermediate, HI, is formed, before the second fast step gives the products. Don't worry if you can't work out what the intermediate is, as it can be rather obscure. You will only be asked very obvious rate-determining steps. Just concentrate on knowing how many molecules of each intermediate are involved in the rds.

Quick check questions

1 The rate of dissociation of the gas SO_2Cl_2 follows this rate law:
rate = $k[SO_2Cl_2]$. If the initial rate is 1.35×10^{-4} mol dm^{-3} s^{-1} and the initial concentration of SO_2Cl_2 is 0.45 mol dm^{-3}, calculate the value of the rate constant *k*.

2 Suggest what the rate-determining step is in this reaction:
$CH_3COCH_3\ (aq) + I_2(aq) \rightleftharpoons CH_3COCH_2I(aq) + HI(aq)$
rate = $k[CH_3COCH_3][H^+][I_2]^0$

How Fast? Concentration–time graphs

When a reaction is taking place, it is possible to measure the concentration of a reactant over time. The *shape* of this *concentration–time graph* tells you the *order* with respect to the reactant being measured.

The following graphs show you the shape which each different order for a reactant gives. You must know these graphs – you can tell the order of a reactant just by looking at them. They all assume that there is one reactant only, A.

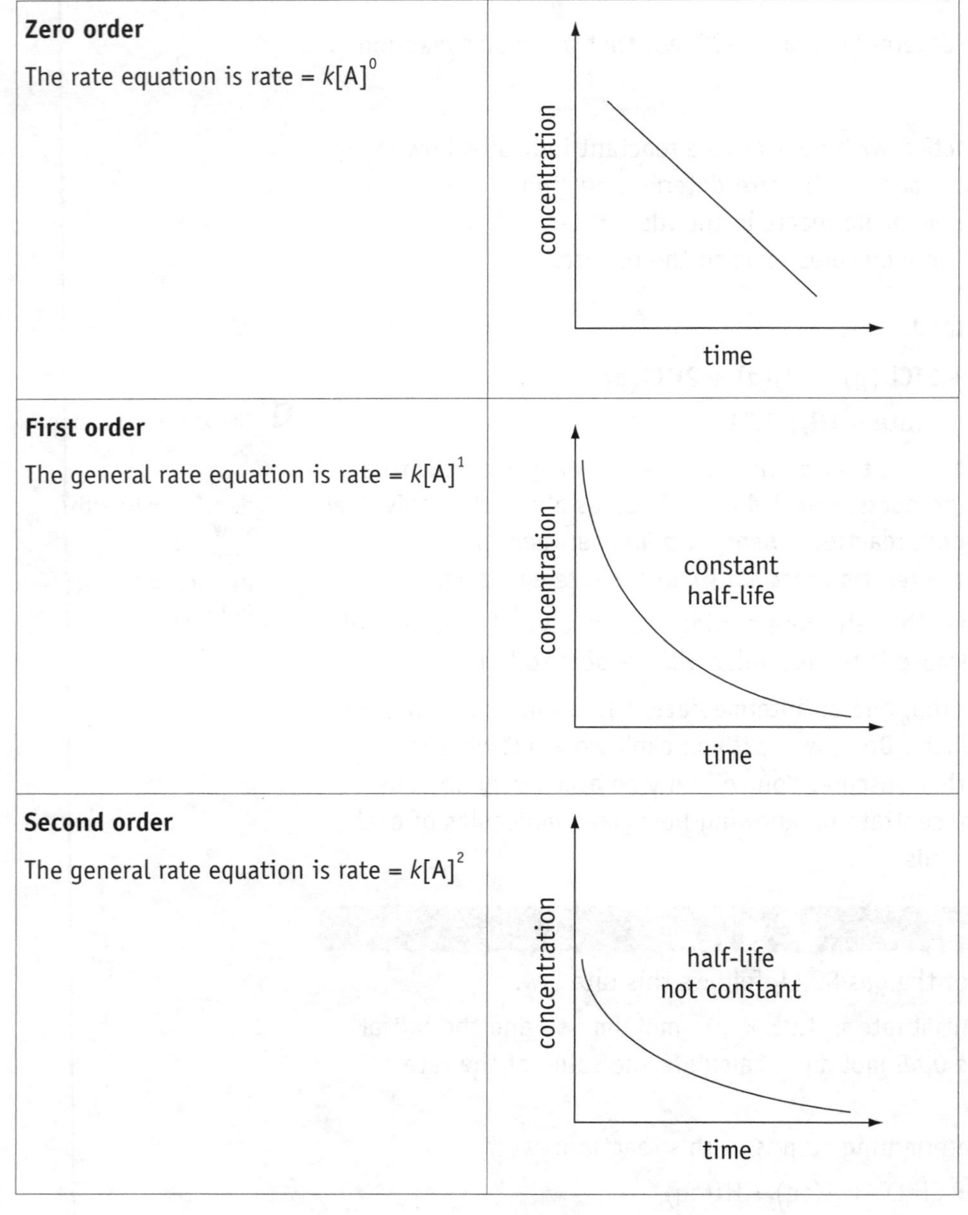

Zero order The rate equation is rate = $k[A]^0$	
First order The general rate equation is rate = $k[A]^1$	
Second order The general rate equation is rate = $k[A]^2$	

The reaction rate at a particular concentration is the tangent to the concentration/time graph.

Reaction half-life

The half-life ($t_{½}$) of a reaction is the time required to reach half the initial concentration.

The units of half-lives depend on the speed of the reaction. They can be in seconds, minutes, hours, days or years. For example, the decomposition of the gas N_2O_5 has $t_{½}$ = 24.0 min at 45°C. So if we start with 0.06 mol dm^{-3} N_2O_5, after 24 min (one half-life) 0.03 mol dm^{-3} remain so 0.03 mol dm^{-3} has decomposed. After another 24 min, 0.015 mol dm^{-3} remain and 0.015 mol dm^{-3} has decomposed, and so on.

Half-life of a first-order reaction

Under fixed conditions the half-life of a first-order reaction is a constant.

$t_{½}$ is a constant for a first-order reaction, no matter what the concentration of the reactant is.

Look at the concentration–time graph for a first-order reaction. This shape is called an *exponential curve*. Exponential curves always have a constant half-life.

How to calculate the half-life – see the curve opposite:

Step 1: Choose any convenient concentration value (one which can be easily halved!) – here it is 32 mol dm^{-3}.

Step 2: Draw a horizontal line to the curve and then a perpendicular down to the time axis.

Step 3: Find half the initial concentration value – here it is 16 mol dm^{-3} – and find the time again. You now have two time values, and the difference between them is the half-life, $t_{½}$. Here it is 10 sec.

Step 4: Find half the concentration again – here it is 8 mol dm^{-3} – and repeat the whole process to find the second half-life. Here it is 10 sec. If the two $t_{½}$ values are the same, the reaction is first order. If they are different, it is not a first-order reaction.

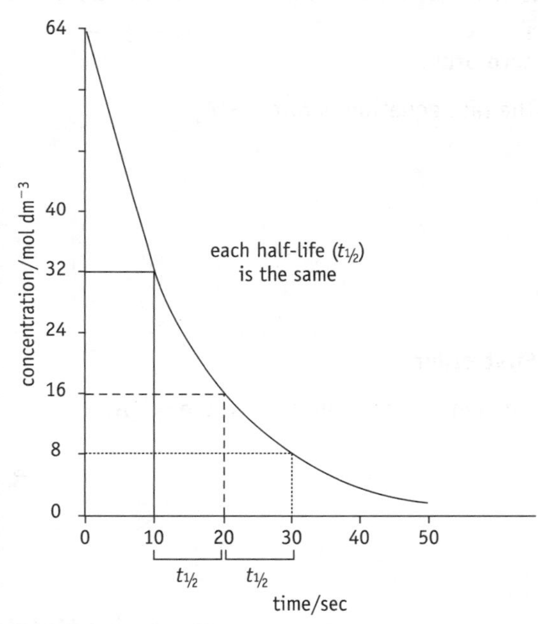

Quick check question

1 Determine whether the following reaction is first order for bromine:

$Br_2(aq) + HCOOH(aq) \rightarrow 2Br^-(aq) + 2H^+(aq) + CO_2(g)$

$[Br_2]$ / mol dm^{-3}	time / min
0.0111	0
0.00811	60
0.0066	120
0.0044	240
0.0020	480
0.0013	600

How Fast?
Rate–concentration graphs

Look at the concentration–time graphs on the previous page. It can be difficult to tell the difference between a first-order and a second-order curve – the second-order curve is steeper, and does not have a constant $t_{1/2}$, but these differences can be difficult to spot. To tell the difference more clearly, we plot *rate–concentration graphs*.

Rate–concentration graphs are obtained by **the initial rates method.** This means that the initial rate is measured in several experiments in which the reactants have different concentrations. Often one concentration remains constant while the others are changed.

It is often more convenient to use the initial rates method than to take concentration and time measurements, especially if it's a slow reaction!

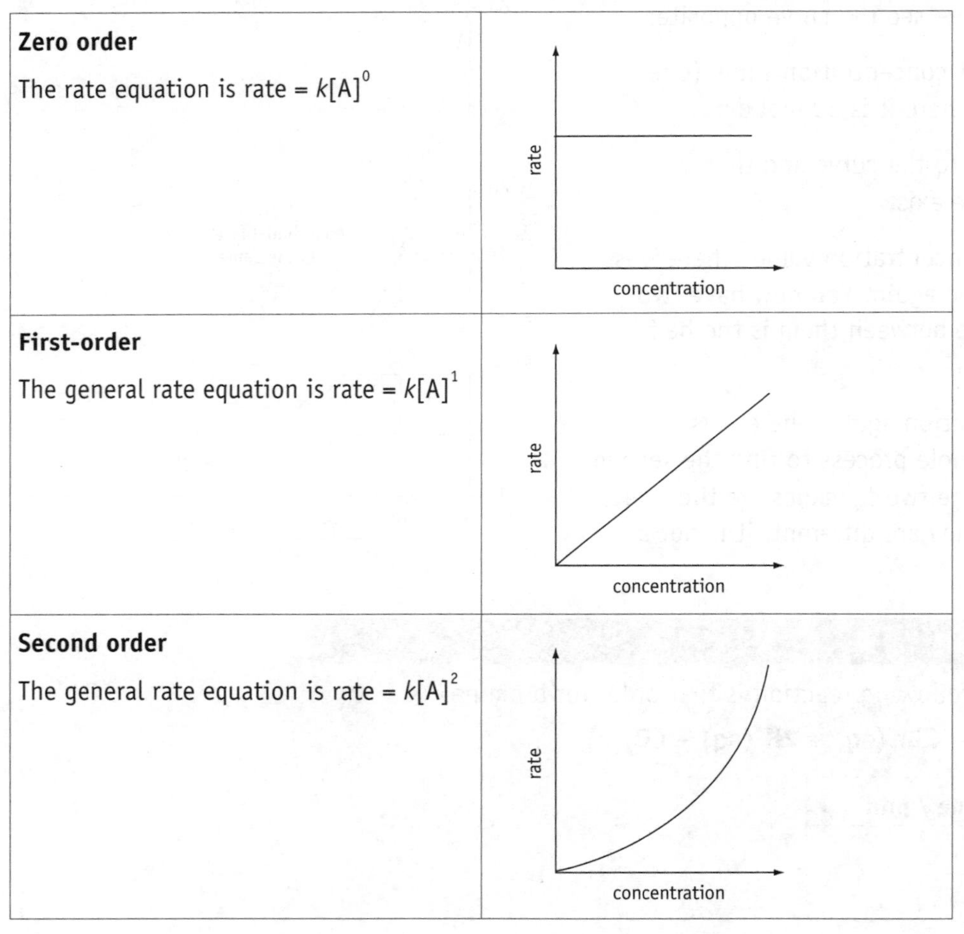

Zero order The rate equation is rate = $k[A]^0$	
First-order The general rate equation is rate = $k[A]^1$	
Second order The general rate equation is rate = $k[A]^2$	

If you plot $(\text{concentration})^2$ against time you get a straight line for a second-order reaction.

✓ *Quick check 1*

How to calculate the order of a reaction

There are different ways of doing this, which depend on the information you are given. Follow this map to work out the order:

✓ Quick check 2

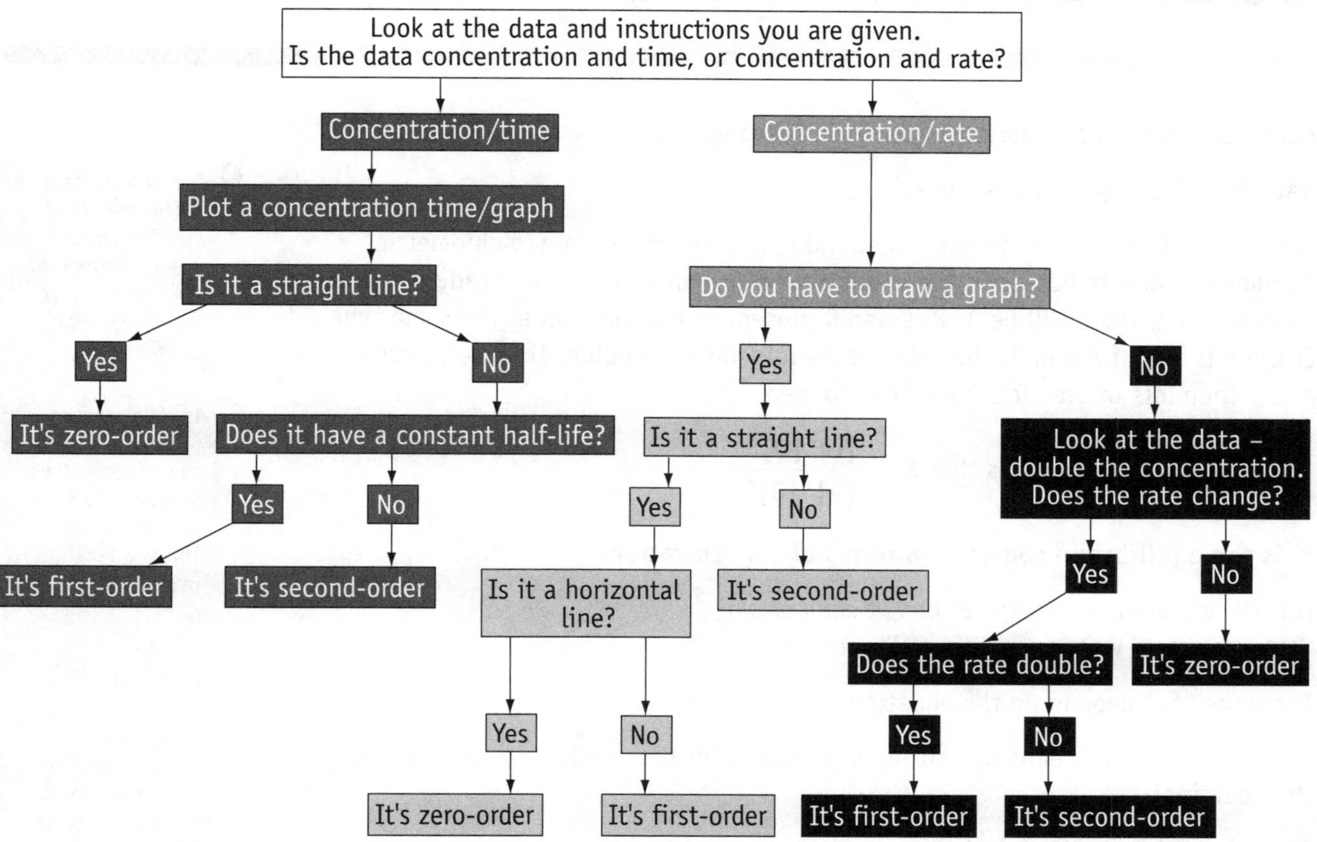

Quick check questions

1 Describe the initial rates method. Sketch the graph you would obtain for a second-order reaction using this method.

2 Calculate the order of reaction for N_2O_5 in the following reaction:

$$2N_2O_5\,(g) \rightarrow 4NO_2(g) + O_2(g)$$

initial rate × 10^{-5} / mol dm^{-3} s^{-1}	initial $[N_2O_5]$ / mol dm^{-3}
3.15	3.00
2.64	2.51
1.18	1.12
1.53	0.50

How Far? The equilibrium law and K_c

Consider a general equilibrium in aqueous solution:

$a\text{A(aq)} + b\text{B(aq)} \rightleftharpoons c\text{C(aq)} + d\text{D(aq)}$

where A, B, C and D are the reactants, and *a*, *b*, *c* and *d* are the stoichiometric number of moles indicated in the equation. At any time, once the **steady state** has been reached, there will be A, B, C and D present in the reaction at the same time. If there is lots of A and B, then the reaction is not proceeding. If there is lots of C and D then lots of products have been formed.

The **equilibrium law** states that $K_c = \dfrac{[\text{C}]^c[\text{D}]^d}{[\text{A}]^a[\text{B}]^b}$

K_c is the equilibrium constant in terms of concentration.

[C], [D] etc are the concentrations *at equilibrium*.
Often these are written as $[\text{C}]^c_{\text{eq}}$, $[\text{D}]^d_{\text{eq}}$ etc.

The *units of* K_c depend on the equation.

Let's see what this means by writing an equation for the equilibrium constant for this reaction:

$2SO_2(g) + O_2(g) \rightleftharpoons 2SO_3(g)$

We simply write the concentration of the product, $[SO_3]$, and raise it to the power of the number of moles shown in the equation, which is two. Divide this by the concentration of the reactants raised to the power of their number of moles:

$$K_c = \frac{[SO_3]^2}{[SO_2]^2[O_2]}$$

The **steady state** means that at equilibrium, *the rate of the forward reaction equals the rate of the back reaction.*

In the expression for K_c the concentrations are all equilibrium concentrations. This is often written as $K_c = \dfrac{[\text{C(aq)}]^c_{\text{eq}}[\text{D(aq)}]^d_{\text{eq}}}{[\text{A(aq)}]^a_{\text{eq}}[\text{B(aq)}]^b_{\text{eq}}}$ In your work, K_c always refers to aqueous solutions at equilibrium, so these symbols are left out to make things clearer.

The units of K_c

What are the *units of* K_c? Well, the units of concentration are mol dm^{-3}, so substitute mol dm^{-3} in the equation for K_c:

$$\text{If } K_c = \frac{[SO_3]^2}{[SO_2]^2[O_2]}$$

$$\text{units} = \frac{(\text{mol dm}^{-3})^2}{(\text{mol dm}^{-3})^2(\text{mol dm}^{-3})}$$

$$\text{units} = \frac{1}{\text{mol dm}^{-3}}$$

$$= \mathbf{dm^3\ mol^{-1}}$$

Notice how $\dfrac{1}{\text{mol dm}^{-3}}$ is written as $dm^3\ mol^{-1}$. Conventionally the unit with the negative power goes last, but you can write this any way you like.

The important thing about the units of K_c is that they are not always the same. Let's look at this reaction:

$$H_2(g) + CO_2(g) \rightleftharpoons CO(g) + H_2O(g)$$

The equation for K_c is: $K_c = \frac{[CO][H_2O]}{[H_2][CO_2]}$

so the units are $\frac{(\text{mol dm}^{-3})(\text{mol dm}^{-3})}{(\text{mol dm}^{-3})(\text{mol dm}^{-3})}$

which cancels out to **no units.**

You can see how the units for K_c depend on the equation for K_c, which in turn depends on the stoichiometric equation. So always make sure you are using the correct stoichiometric equation when you write the expression for K_c.

Important facts about K_c

A large value of K_c means a high yield of products.

A small value of K_c means that a high proportion of reactants are present.

Most organic reactions are equilibrium reactions.

The effect on K_c when conditions are changed

Students often get confused between the change in the *position of the equilibrium* determined by Le Chatelier's principle, and K_c.

The main point to remember here is that the value of K_c (or K_p, see page 76) *only changes if the temperature is changed*. If the *concentration or pressure* is changed, the *position* of the equilibrium may change – remember Le Chatelier's principle in How Far, How Fast? – but the *value of K_c* stays the same.

Le Chatelier's principle states that if an equilibrium is disturbed, it changes position to oppose the disturbance.

What happens to K_c when the temperature is changed?

This depends on whether the equilibrium is exothermic or endothermic.

The terms *exothermic* and *endothermic* used here refer to the *forward reaction.*

	exothermic reaction	**endothermic reaction**
temperature raised	K_c decreases	K_c increases
temperature reduced	K_c increases	K_c decreases

Quick check question

1 Write expressions for K_c ,including units, for the following equilibria:

(a) $PCl_5(g) \rightleftharpoons PCl_3(g) + Cl_2(g)$

(b) $C_2H_4(g) + H_2O(g) \rightleftharpoons C_2H_5OH(g)$

(c) $H_2(g) + Br_2(g) \rightleftharpoons 2HBr(g)$

How Far? How to calculate the value of K_c

You need to be able to calculate the numerical value of K_c. There are two different types of approach, depending on the information you are given in the question.

1 If you are given the *equilibrium* concentrations

The equation for K_c uses the concentrations at equilibrium. If you are given these, simply substitute into the K_c equation:

Worked example

Calculate K_c for

$$H_2(g) + CO_2(g) \rightleftharpoons CO(g) + H_2O(g)$$

given $[H_2]_{eq} = 0.53$ mol dm^{-3} $[CO_2]_{eq} = 80.53$ mol dm^{-3}
$[CO]_{eq} = 9.47$ mol dm^{-3} $[H_2O]_{eq} = 9.47$ mol dm^{-3}

Step 1: Write out the expression for K_c from the stoichiometric equation.

$$K_c = \frac{[CO][H_2O]}{[H_2][CO_2]}$$

Step 2: Substitute with the equilibrium concentrations you are given.

$$K_c = \frac{[CO][H_2O]}{[H_2][CO_2]} = \frac{9.47 \times 9.47}{0.53 \times 80.53}$$

$$= 2.10 \text{ (no units)}$$

Imagine you were **not** given $[H_2O]_{eq}$. What would you do? Well, the stoichiometric equation tells you that at equilibrium the number of moles of H_2O and the number of moles of CO are the same. So if $[CO] = 9.47$ mol dm^{-3} then $[H_2O] = 9.47$ mol dm^{-3}

2 If you are given the *initial* concentrations

What if the equilibrium concentrations of the reactants are unknown? You can calculate them from *initial* amounts of the reactants and the *equilibrium concentration of the product.*

If you know how many moles of product are present *at equilibrium*, the stoichiometric equation will tell you how many moles of reactant are present *at equilibrium.*

Worked example

Calculate K_c for the equilibrium

$$2SO_3(g) \rightleftharpoons 2SO_2(g) + O_2(g)$$

if 0.067 moles of SO_3 was introduced into the equilibrium vessel (volume 1.52 dm^3) and 0.0142 moles of SO_2 was found at equilibrium.

Step 1: Construct the equation for K_c:

$$K_c = \frac{[SO_2]^2[O_2]}{[SO_3]^2}$$

Step 2: Make up a table based on the stoichiometric equation. You are going to need the *initial number of moles* and the *equilibrium number of moles*. Finally you are going to need *equilibrium concentrations in mol dm^{-3}*.

	$2SO_3$	$\rightleftharpoons$	$2SO_2$ +	O_2
initial amounts/mol				
equilibrium amounts/mol				
equilibrium concentration/mol dm^{-3}				

Step 3: Put the values you know under the appropriate species in the equation.

	$2SO_3$	$\rightleftharpoons$	$2SO_2$ +	O_2
initial amounts/mol	**0.067**		**0**	**0**
equilibrium amounts/mol			**0.0142**	
equilibrium concentration/mol dm^{-3}				

Step 4: Now you can put in the number of moles of the other species:

	$2SO_3$	$\rightleftharpoons$	$2SO_2$ +	O_2
initial amounts/mol	**0.067**		**0**	**0**
equilibrium amounts/mol	**0.067 – 0.0142 +0.0071 = 0.0457**		**0.0142**	**0.0142/2 = 0.0071** (must be ½ moles of SO_2)
equilibrium concentration/mol dm^{-3}				

Step 5: Now you can work out the concentration by dividing the number of moles by the volume.

	$2SO_3$	$\rightleftharpoons$	$2SO_2$ +	O_2
initial amounts/mol	**0.02**		**0**	**0**
equilibrium amounts/mol	**0.0457**		**0.0142**	**0.0142/2 = 0.0071** (must be ½ moles of SO_2)
equilibrium concentration/mol dm^{-3}	**0.0457/1.52 = 0.0301 mol dm^{-3}**		**0.0142/1.52 = 0.00934 mol dm^{-3}**	**0.0071/1.52 = 0.00467 mol dm^{-3}**

Step 6: Substitute these values into the expression for K_c.

$$K_c = \frac{[SO_2]^2[O_2]}{[SO_3]^2} = \frac{0.00934^2 \times 0.00467}{0.0301^2} = 2.1 \times 10^{-4}\ \text{mol dm}^{-3}$$

Quick check question

1 0.204 mol $CH_3COOC_2H_5$ and 0.645 mol H_2O were introduced into a flask, acidified and left until a steady state was reached. A sample was then analysed, and 0.114 mol of CH_3COOH was found. Calculate K_c for the reaction

$$CH_3COOC_2H_5(l) + H_2O(l) \rightleftharpoons CH_3COOH(l) + C_2H_5OH(l)$$

How Far? The equilibrium law and K_p

So far we have been studying equilibria where the reactants and products are in aqueous solution. For equilibria where all the species are gases, it is often more convenient to measure *pressure* than concentration.

The equilibrium constant can be expressed in terms of *partial pressures of the gases.* Instead of K_c we use $\boldsymbol{K_p}$.

What is partial pressure?

The partial pressure p of a gas in a mixture is the pressure of that gas alone in the same volume as the mixture.

- **How is the partial pressure related to the total or overall pressure of the mixture of gases?**

The **total pressure p_{total}** of a mixture of gases = **sum of the partial pressures** exerted by each of the gases in the mixture.

We say that $\boldsymbol{p_{total} = p_1 + p_2 + p_3 + \ldots}$ where p_1, p_2 etc are the partial pressures of the individual gases.

- **How do we know what the partial pressure is?**

If there are two gases, A and B, then

$\boldsymbol{p_A}$ **= mole fraction of A** $\times$ $\boldsymbol{p_{total}}$

and **the mole fraction of** $A = \dfrac{n_A}{n_A + n_B}$

where n_A is the number of moles of A, and n_B is the number of moles of B.

This means that $p_A = \dfrac{n_A}{n_A + n_B} \times p_{total}$ and $p_B = \dfrac{n_B}{n_A + n_B} \times p_{total}$

The equilibrium constant in terms of partial pressure is called $\boldsymbol{K_p}$. *It has exactly the same form as K_c.*

Instead of concentrations, partial pressures are used. So for the reaction
$a\text{A} + b\text{B} \rightleftharpoons c\text{C} + d\text{D}$

$$K_p = \frac{p_C^c \times p_D^d}{p_A^a \times p_B^b}$$

Units of K_p

The units of p are kiloPascals, kPa. Work out the units of K_p in Pa or kPa in the same way that you would for K_c in mol dm^{-3}.

Worked example

State the units of the equilibrium constant K_p for this reaction:

$$PCl_3(g) + Cl_2(g) \rightleftharpoons PCl_5(g)$$

Step 1: Write the expression for K_p.

$$K_p = \frac{p_{PCl_5}}{p_{PCl_3} \times p_{Cl_2}}$$

Step 2: Substitute the partial pressures with Pascals.

$$K_p = \frac{Pa}{Pa \times Pa} = \frac{1}{Pa} = Pa^{-1}$$

> Look carefully at the units for pressure you are given. Usually gas pressures are given in kPa, not Pa. (1kPa = 10^3 Pa). You can use kPa instead of Pa in the K_p equation – see the worked example below.

How to calculate K_p

You use exactly the same method to calculate K_p as you did to calculate K_c. But there are two extra steps for gaseous reactions – you have to work out the *mole fractions* and then the *partial pressures* before you can substitute into the K_p equation.

Worked example

In the equilibrium $N_2O_4(g) \rightleftharpoons 2NO_2(g)$ there were 0.1 moles of N_2O_4 and 0.4 moles of NO_2. The total pressure was 100 kPa. Calculate the value of K_p for this equilibrium.

Step 1: Write the expression for K_p.

$$K_p = \frac{pNO_2{}^2}{p_{N_2O_4}}$$

Step 2: Calculate the mole fraction of each gas.

$$\textbf{mole fraction of } N_2O_4 = \frac{\text{no. moles } N_2O_4}{\text{total no. of moles}} = \frac{0.1}{0.5} = 0.2 \text{ mol}$$

$$\textbf{mole fraction of } NO_2 = \frac{\text{no. moles } N_2O_4}{\text{total no. of moles}} = \frac{0.4}{0.5} = 0.8 \text{ mol}$$

Step 3: Calculate the partial pressure of each gas.

$$p_{N_2O_4} = 0.2 \times 100 \text{ kPa and } p_{NO_2} = 0.8 \times 100 \text{ kPa}$$

Step 4: Put the values of the partial pressures into the expression for K_p, and calculate.

$$K_p = \frac{p_{NO_2}}{p_{N_2O_4}} = \frac{(0.8 \times 100)^2}{0.2 \times 100} = 320$$

Step 5: Add the units.

$$\frac{kPa^2}{kPa} = kPa \text{ so } K_p = 320 \text{ kPa}$$

> The same things to remember apply to K_p as K_c:
> - A large value of K_p means a high yield of products.
> - A small value of K_p means that a high proportion of reactants are present.
> - The value of K_p *changes* when the *temperature* is changed.
> - It does *not change* when the *pressure is changed.*

? Quick check question

1 The gas N_2O_4 was allowed to come to equilibrium with NO_2 in a closed flask, so that the total pressure was 2.14 kPa. The amounts of gas at equilibrium were determined as 0.021 mol N_2O_4 and 0.006 mol NO_2. Calculate K_p for the equilibrium

$N_2O_4 \rightleftharpoons 2NO_2$.

Acids, bases and buffers

The chemistry of acids, bases and buffers is an important area. For example, the relative strengths of acids influences the formation of nitronium ions in the nitration of benzene, and the understanding of pH and buffers is essential in biology.

The Bronsted–Lowry theory of acids and bases

The main aspects of this theory state that:

- **Acids are proton (H^+ ion) donors** whilst **bases are proton acceptors.**
- In an acid-base reaction, the acid donates a proton to form a **conjugate base** and the base accepts a proton to form a **conjugate acid.**

✓ Quick check 1

Generally, the reaction between a 'Bronsted–Lowry acid' and a 'Bronsted–Lowry base' may be written as follows:

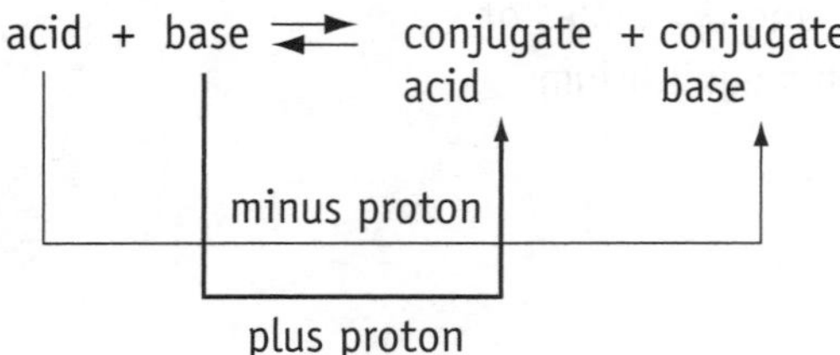

- Bases must always have lone pair electrons to form a dative covalent bond with the proton
- You can understand the concept of conjugate acids and bases if you think of the reverse reaction and think which species would act as the proton donor (conjugate acid) and which would accept the proton (the conjugate base).
- An example is the reaction of hydrogen chloride and water. In this reaction, the hydrogen chloride is the acid and the water is the base. The HCl donates protons more strongly than water and water has lone-pair electrons with which it can form a dative covalent bond with the proton.

$$\underset{\text{acid}}{HCl} + \underset{\text{base}}{H_2O} \rightleftharpoons \underset{\text{conjugate acid}}{H_3O^+} + \underset{\text{conjugate base}}{Cl^-}$$

- The H_3O^+ ion is called the **oxonium** (or hydroxonium) ion and is the real ion found in aqueous solutions of acids. The H^+ ion cannot really exist. It is not stable because it does not have the two electrons necessary for stability.

✓ Quick check 2

- Water can also act as an acid when it is with a stronger proton acceptor such as ammonia.

$$\underset{\text{acid}}{H_2O} + \underset{\text{base}}{NH_3} \rightleftharpoons \underset{\text{conjugate acid}}{NH_4^+} + \underset{\text{conjugate base}}{OH^-}$$

- The strength of an acid depends on its ability to donate protons.

The reaction between concentrated sulphuric acid and concentrated nitric acid is an acid-base because the sulphuric acid is the stronger proton donor.

$$\underset{\text{acid}}{H_2SO_4} + \underset{\text{base}}{HNO_3} \rightleftharpoons \underset{\text{conjugate acid}}{H_2NO_3^+} + \underset{\text{conjugate base}}{HSO_4^-}$$

✓ Quick check 3

- Because of its dual nature, as an acid (able to donate protons) and as a base (able to accept protons), water is said to be **amphoteric**.

Some more examples are given below. Note: the acids and their conjugate bases are underlined.

$\underline{H_2SO_4} + H_2O \rightleftharpoons \underline{HSO_4^-} + H_3O^+$

$\underline{HSO_4^-} + H_2O \rightleftharpoons \underline{SO_4^{2-}} + H_3O^+$

$\underline{HI} + HCl \rightleftharpoons \underline{I^-} + H_2Cl^+$

$\underline{CH_3CO_2H} + H_2O \rightleftharpoons \underline{CH_3CO_2^-} + H_3O$

> NOTE: acids such as H_2SO_4 which have 2 protons that it can donate are called **diprotic** acids.
>
> HCl has one proton to donate and is **monoprotic.**

Why are some acids stronger than others?

A common questions in exams asks you to explain why one compound is a stronger acid than another. It **is not enough** to state that it is a stronger proton donor. There are two possible answers depending on the comparison.

Answer I The bond between the proton and the rest of the molecule or ion is weakened more in one than the other.

Answer II The product for one is more stable than the other.

Examples

a A question often asked in **'Chains and Rings'** papers is why phenol is more acidic than ethanol. The O–H bond in phenol is weakened by the electrons in the bond being pulled into the ring. The O–H bond in ethanol is strengthened because the CH_3CH_2 (ethyl) group pushes electrons away (positive inductive effect) thus strengthening the O–H bond. The overall effect is that the phenol donates its proton more easily than does ethanol.

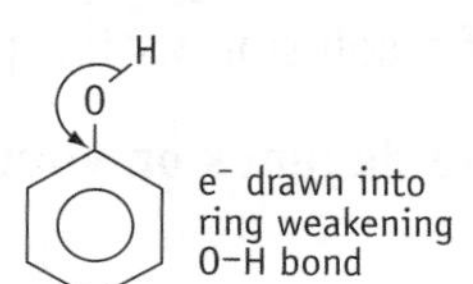

CH_3CH_2–O–H

e⁻ pushed away strengthening O–H bond

An alternative answer is that the phenoxide ion, $C_6H_5O^-$ is stabilised by the delocalisation of the negative charge on the benzene ring. The negative charge on the ethoxide, $C_2H_5O^-$ is not stabilised. The phenoxide ion is therefore more likely to be formed.

b Hydrogen fluoride is a much weaker acid than hydrogen chloride because the H–F bond is much stronger than the H–Cl bond and the HCl more easily donates the proton than does HF. Therefore the reaction between HCl and water happens more easily.

> $HCl + H_2O \rightarrow H_3O^+ + Cl^-$

? *Quick check questions*

1. For the acid–base equilibria shown below, underline the acid and ring its conjugate base.
   (a) $HClO_4 + HNO_3 \rightleftharpoons ClO_4^- + H_2NO_3^+$
   (b) $HBr + H_2O \rightleftharpoons H_3O^+ + Br^-$
   (c) $CH_3NH_2 + H_2O \rightleftharpoons CH_3NH_3^+ + OH^-$
2. (a) What is the name given to the H_3O^+ ion?
   (b) Using two of the reactions in question 1 above explain why water is said to be amphoteric.
3. $HClO_4$ and HNO_3 are both acids in aqueous solution.
   (a) Give the equations for their reactions with water
   (b) Using the relevant equation in question 1a explain which one is the stronger proton donor.

Acids and bases in aqueous solution

The pH scale

NOTE p is shorthand for $-\log_{10}$

The concentrations of aqueous hydrogen ions vary widely and the range would be impossible to represent on a linear scale. To overcome this problem a logarithmic scale is used called the **pH scale**.

On this scale

$$\mathbf{pH = -\log_{10}[H^+(aq)]}$$

Remember: in aqueous solution the term $H^+(aq)$ is used instead of $[H_3O^+(aq)]$

The converse of this is expressing the $[H^+(aq)]$ in terms of the pH.

$$\mathbf{[H^+(aq)] = 10^{-pH}}$$

Quick check 1

To calculate pH you simple have to use your calculator keys correctly. If you can do this then this section isn't too difficult.

NOTE: For some calculators the [log] key is pressed first, then the value you are computing then the [=] key.

Worked examples

1 What is the pH of a solution with a $[H^+(aq)]$ equal to 3×10^{-4} mol dm^{-3}?

The exact procedure depends on your calculator and you should consult your manual.

Typical procedures are as follows:

Step 1: Enter the value 3 x 10^{-4}. The [EXP] or [EE] keys will be needed here.

Step 2: Press [LOG] and then the [ANS] or [=] key and your calculator will give the figure –3.52878 etc.

Step 3: Remember pH = $-\log_{10}$ $[H^+(aq)]$ and therefore your final answer to 3 significant figures (3 sig. fig.) is –(–3.53) which is +3.53.

2 What is the $[H^+(aq)]$ if the pH is 7.2.

To find $[H^+(aq)]$ we use the formula **$[H^+] = 10^{-pH}$.**

Step 1: Find the [10^x] key and press it. **NOTE:** The 10^x key is usually [SHIFT][log]

Step 2: Press 7.2 then [+/–].

Step 3: Press your [ANS] or [=] key and you get the answer 6.31×10^{-8} mol dm^{-3}

Quick check 2

The ionisation of water

$$K = \frac{[H^+(aq)][OH^-(aq)]}{[H_2O(l)]}$$

- Water ionises as follows: $H_2O \rightleftharpoons H^+(aq) + OH^-(aq)$
  The equilibrium constant may therefore be written as shown opposite.
- At 25 °C this equilibrium lies well to the left and the equlibrium concentrations of $[H^+(aq)]$ and $[OH^-(aq)]$ are both equal to 10^{-7} mol dm^{-3}. Therefore $[H_2O(l)]_{eqm}$ is virtually unchanged from that of undissociated water and can be treated as constant. A new equilibrium constant K_w is now used and this can be represented by the following equation:

REMEMBER THIS! $K_w = [H^+(aq)]_{eqm}[OH^-(aq)]_{eqm}$ $mol^2\ dm^{-6}$

- At 25 °C $K_w = 1 \times 10^{-14}$ $mol^2\ dm^{-6}$
  Since $[H^+(aq)] = [OH^-(aq)]$ then $K_w = [H^+(aq)]^2$
  $[H^+(aq)]^2 = 1 \times 10^{-14}$ $mol^2\ dm^{-6}$
  $[H^+(aq)] = \sqrt{1 \times 10^{-14}}$ $mol^2\ dm^{-6}$ $= 1 \times 10^{-7}$ $mol\ dm^{-3}$
- When we work out the pH for this $[H^+(aq)]$ we get a value of 7, and this explains why for all these years you have associated this value with a neutral pH.

When you work out the pH for 1×10^{-7} $mol\ dm^{-3}$ you must enter 1 then [EXP] then [+/–] then 7.

The pH values of aqueous solutions

This is a section where we have to assume you know the expression for pH and once again you must use your calculator.

The pH values of solutions of strong acids

- The important point here is that strong acids are fully ionised and therefore we can assume 100% ionisation.
- For example 0.1 mol dm^{-3} HCl ionises as follows:

$$\mathbf{HCl(aq) \rightarrow H^+(aq) + Cl^-(aq)}$$

- For 100% ionisation every HCl gives one $H^+(aq)$
  $\therefore [H^+(aq)] = [HCl(aq)] = 0.1$ mol dm^{-3}
  $\therefore$ pH $= -\log_{10}0.1 = \underline{1}$

The pH values of solutions of strong bases

- For strong bases the assumption of 100% ionisation is still valid.
- For example NaOH ionises as follows:

$$\mathbf{NaOH(aq) \rightarrow Na^+(aq) + OH^-(aq)}$$

- Therefore, if we have a solution of 0.02 mol dm^{-3} NaOH then every NaOH gives a $OH^-(aq)$ ion

$$\mathbf{[OH^-(aq)] = [NaOH(aq)] = 0.02\ mol\ dm^{-3}.}$$

- *But pH is calculated on the basis of $[H^+(aq)]$* and therefore we have to find $[H^+(aq)]$ using our value of $[OH^-(aq)]$. To do this we use the relationship:

$$\mathbf{K_w = [H^+(aq)][OH^-(aq)] = 1 \times 10^{-14}\ mol^2\ dm^{-6}}$$

Substituting in our value for $[OH^-(aq)]$ we have:
$1 \times 10^{-14} = [H^+(aq)] \times 0.02$
$[H^+(aq)] = 1 \times 10^{-14}/0.02 = 5 \times 10^{-13}$ mol dm^{-3}
$\therefore$ pH $= -\log_{10} 5 \times 10^{-13} = \underline{12.3}$

Quick check 3

Does this look right? Yes it does, because if you remember your GCSE Chemistry, alkalis had pH values greater than 7

? Quick check questions

1 Calculate the pH values of solutions with the following $[H^+(aq)]$ values:
(a) 0.01 mol dm^{-3} (b) 2×10^{-9} mol dm^{-3} (c) 3×10^{-4} mol dm^{-3}

2 Calculate the concentration of $H^+(aq)$ ions for solutions with the following pH values:
(a) 2.5 (b) 7.4 (c) 10.5 (d) 13.6

3 Calculate the pH values of the following solutions:
(a) HCl (i) 0.010 mol dm^{-3} (ii) 3×10^{-5} mol dm^{-3}
(b) NaOH (i) 0.010 mol dm^{-3} (ii) 3×10^{-5} mol dm^{-3}

The chemistry of weak acids

- Monobasic (monoprotic) acids (one proton produced per molecule) can be represented by HA.
- Weak acids are **poor proton donors** and are only **partially ionised** in aqueous solution. Therefore the dissociation is not 100% and $[H^+(aq)]$ cannot be calculated directly from the concentration of the acid.
- The equilibrium between a weak acid and its constituent ions in aqueous solution is generally written:

$$\mathbf{HA(aq) \rightleftharpoons H^+(aq) + A^-(aq)}$$

- The equilibrium constant for this reaction is called the acid dissociation constant (**K_a**). The expression for K_a is shown opposite.

The magnitude of K_a is a measure of how much the acid ionises. Therefore the higher the value of K_a the stronger the acid.

$$K_a = \frac{[H^+(aq)]_{eqm}[A^-(aq)]_{eqm}}{[HA(aq)]_{eqm}} \text{ mol dm}^{-3}$$

The meaning of the term pK_a

pK_a = $-\log_{10}K_a$. For the example above $K_a = 4.8 \times 10^{-5}$ mol dm^{-3}.

pK_a = $-\log_{10}4.8 \times 10^{-5} = 4.32$.

RULE The higher the value of pK_a the weaker the acid.

NOTE: The units of K_a are mol dm^{-3} because the units in the expression are $\frac{\text{conc.} \times \text{conc.}}{\text{conc.}}$ = conc. and the units of concentration are mol dm^{-3}.

Worked example

Which is the stronger of the two acids, ethanoic acid ($K_a = 1.7 \times 10^{-5}$ mol dm^{-3}) or chloric(I) acid ($K_a = 3.7 \times 10^{-8}$ mol dm^{-3})?

Ethanoic acid pK_a = $-\log_{10}1.7 \times 10^{-5} = 4$.
Chloric(I) acid pK_a = $-\log_{10}3.7 \times 10^{-8} = 7.43$

The weaker of the two acids is therefore chloric(I) acid.

NOTE this is the opposite way round to the rule for K_a.

Finding K_a from pK_a

As with pH the value of K_a can be deduced from pK_a by using the relationship $K_a = 10^{-pK_a}$.

Worked example

The pK_a of phenol is 9.9. What is its K_a?

$K_a = 10^{-pK_a} = 10^{-9.9} = \underline{1.26 \times 10^{-10}}$.

There are two types of question which involve K_a and $[H^+(aq)]$.

Type 1 Finding K_a from pH

Worked example

What is K_a for a weak acid if a solution of the acid with concentration 0.01 mol dm^{-3} has a pH of 5.5?

Step 1: Calculate $[H^+(aq)]$. In this case the pH is 5.5 *

$\therefore$ $[H^+(aq)] = 10^{-5.5} = 3.16 \times 10^{-6}$ mol dm^{-3}

* If this was a strong acid and 100% ionised $[H^+(aq)]$ would be 0.01 mol dm^{-3} and the pH would be 2.

Step 2: Since there are equal numbers of $H^+(aq)$ and $A^-(aq)$ produced in the ionisation we can assume that **$[H^+(aq)] = [A^-(aq)]$.**

$\therefore [H^+(aq)][A^-(aq)] = [H^+(aq)]^2$
In this case $[H^+(aq)]^2 = 9.99 \times 10^{-12}$

Step 3: Here we make the assumption that **[HA(aq)] at equilibrium is virtually the same as it was at the start** because as a weak acid so little of HA ionises.
$[HA(aq)]_{eqm} = [HA(aq)]_{start} = 0.01\ mol\ dm^{-3}$

Step 4: Substitue these values into the expression for K_a and we have:

$$K_a = \frac{9.99 \times 10^{-12}}{0.01} = \underline{9.99 \times 10^{-10}}\ mol\ \ dm^{-3}$$

NOTE: The low value for K_a emphasises that this is a weak acid.

✓ Quick check 2

Type 2 Finding pH from K_a

Worked example

The value of K_a for a weak acid is 4.8×10^{-5} mol dm–3. What is the pH of a solution of the acid with a concentration 0.1 mol dm^{-3}?

Step 1: Make the same assumptions as before, i.e. $[H^+(aq)] = [A^-(aq)]$ and $[HA(aq)]_{eqm} = [HA(aq)]_{start}$

$\therefore K_a = [H^+(aq)]^2/[HA(aq)]_{start}$
$\therefore [H^+(aq)]^2 = K_a \times [HA(aq)]_{start}$

Step 2: Substitute in these values
$\therefore [H^+(aq)]^2 = 4.8 \times 10^{-5} \times 0.1 = 4.8 \times 10^{-6}$
$\therefore [H^+(aq)] = \sqrt{}(4.8 \times 10^{-6}) = 2.19 \times 10^{-3}\ mol\ dm^{-3}$

Step 3: Calculate pH from $[H^+(aq)]$. pH = $-\log_{10}2.19 \times 10^{-3} = \underline{2.66}$

NOTE: The question may give you the pK_a and not K_a and ask you to find the pH. If this is the case simply calculate K_a using $K_a = 10^{-pK_a}$ and carry on as in the example.

✓ Quick check 3

? Quick check questions

1 (a) The K_a values of two weak acids HA and HB are as follows:
K_a (HA) = 2.9×10^{-3} K_a (HB) = 4.5×10^{-6}
(a) Calculate pK_a of (i) HA and (ii) HB.
(b) Which is the stronger of the two acids?
(c) Write an equation for the reaction between HA and HB (refer to the Bronsted–Lowry theory)

2 The pH of a 0.01 mol dm^{-3} solution of butanoic acid is 3.41. What is the value of K_a of the acid?

3 The pK_a of silicic acid is 9.9.
(a) What is the value of K_a?
(b) What is the pH of a 0.001 mol dm^{-3} solution of the acid?

Buffers – how they work and their pH

A buffer solution is one which resists changes in pH. Buffers are important solutions, especially in biological systems. A good understanding of how they work is essential if you are to use them properly.

How buffer solutions work

The two main types of buffer solution are

1 Solutions of weak acids and their ions (conjugate bases), for example ethanoic acid and ethanoate ions (CH_3COOH/CH_3COO^-).

In this first example, the CH_3COO^- ions are supplied by a solution of the sodium salt, sodium ethanoate which is fully dissociated.

This type of buffer will give a pH less than 7.

2 Solutions of weak bases and their ions (conjugate acids), for example ammonia and ammonium ions (NH_3/NH_4^+).

In this second example the NH_4^+ ions come from ammonium chloride.

This type of buffer will give a pH greater than 7.

How do buffer solutions maintain a constant pH?

Example Ethanoic acid and ethanoate ions

The equilibrium is: $CH_3COOH(aq) \rightleftharpoons H^+(aq) + CH_3COO^-(aq)$

- Extra $H^+(aq)$ ions are removed by combination with the conjugate base $CH_3COO^-(aq)$ ions to form largely undissociated CH_3COOH (only 4 out of every 1000 ionise). Therefore the $[H^+(aq)]$ and hence the pH remain constant. Note that this is possible because of the high concentration of $CH_3COO^-(aq)$ ions provided by the complete dissociation of the sodium ethanoate.
- If extra $OH^-(aq)$ ions are added they react with $H^+(aq)$ ions (neutralisation) to form water thus disturbing the equilibrium. The CH_3COOH dissociates until K_a is restored to its proper value and therefore the $[H^+(aq)]$, and hence the pH, is back to its previous value.
- Therefore the stable pH of the acidic buffer is maintained by:
  - a high $[CH_3COO^-(aq)]$ which mops up added $H^+(aq)$
  - a high $[CH_3COOH(aq)]$ which supplies $H^+(aq)$ to mop up added $OH^-(aq)$ ions

Remember the neutralisation reaction is $H^+(aq) + OH^-(aq) \rightarrow H_2O(l)$

✓ Quick check 1

• Calculating the pH of a buffer solution

The equation used is a rearrangement of the equation for the acid dissociation constant of a weak acid.

$$[H^+(aq)] = K_a \times \frac{[acid]}{[salt]}$$

Worked example

What is the pH of a solution, 0.02 mol dm^{-3} with respect to ethanoic acid ($K_a = 1.7 \times 10^{-5}$ mol dm^{-3}) and 0.05 mol dm^{-3} with respect to sodium ethanoate?

Step 1: Substituting in the values for K_a, [acid] and [salt] into the equation we have:

$$[H^+(aq)] = 1.7 \times 10^{-5} \times (0.02/0.05) = 6.8 \times 10^{-6} \text{ mol dm}^{-3}$$

Step 2: Calculate pH from $[H^+(aq)]$

$$pH = -\log_{10} 6.8 \times 10^{-6} = \underline{5.17}$$

An alternative equation for the pH of a buffer is:

$$pH = pK_a + \log_{10}\frac{[salt]}{[acid]}$$

Worked example

What is the pH of a solution containing 100 cm^3 of 0.01 mol dm^{-3} ethanoic acid ($K_a = 1.7 \times 10^{-5}$ mol dm^{-3}) and 200 cm^3 of 0.02 mol dm^{-3} sodium ethanoate?

Step 1: Calculate the final concentrations of the ethanoic acid and sodium ethanoate.
The final volume is 300 cm^3.

∴ The concentration of the ethanoic acid = (100/300) × 0.01
= 0.00333 mol dm^{-3}

The concentration of sodium ethanoate = (200/300) × 0.02 = 0.0133

Step 2: $\frac{[salt]}{[acid]} = (0.0133/0.0033) = 4.00$

$$\therefore \log_{10}\frac{[salt]}{[acid]} = \log_{10} 4 = 0.602$$

Step 3: $pK_a = -\log_{10} K_a = 4.77$

Step 4: Substituting these values in the equation

$$pH = 4.77 + 0.602 = \underline{5.30}$$

✓ *Quick check 2*

? *Quick check questions*

1 Explain how a solution of sodium ethanoate and ethanoic acid can function as a buffer.

2 A solution of 100 cm^3 of hydrofluoric acid (0.03 mol dm^{-3}) and 50 cm^3 of sodium fluoride (0.03 mol dm^{-3}) were mixed together.
(K_a of hydrofluoric acid = 5.6×10^{-4} mol dm^{-3}.

(a) What is the final concentration of the acid?

(b) What is the final concentration of the sodium fluoride?

(c) Calculate the pH of the buffer solution formed.

(d) Explain how you think a mixture of HF and NaF may act as a buffer solution.

The importance of buffers in the body

- For enzymes and other important compounds to function efficiently in the body, the pH of the blood should not deviate too far from the value of 7.4. For example when the $[H^+]$ rises (and pH falls) the equilibrium below is moved to the right.

  $$\mathbf{HbO_2 + H^+(aq) \rightleftharpoons HbH^+ + O_2}\ \textbf{(Hb = haemoglobin)}$$

  This means that the haemoglobin does not carry oxygen round the body and aerobic respiration cannot take place. This happens when vigorous **anaerobic** exercise takes place. A lot of lactic acid is formed, lowering the pH, and this leads to less oxygen being carried around the body. This explains why we cannot sprint very quickly for long distances.

- When carbon dioxide is produced in respiration, protons are formed when the carbon dioxide dissolves and produces a more acid solution.

  $$\mathbf{CO_2(aq) + H_2O(l) \rightleftharpoons H^+(aq) + HCO_3^-(aq)}$$

  In fact, the above reaction forms the basis of the most important buffer in the blood.

  - If $[H^+(aq)]$ rises (and the pH falls), the above equilibrium is disturbed. To maintain the equilibrium constant, K_a, of the reaction at a constant value, the extra $H^+(aq)$ ions are removed by reaction with the $HCO_3^-(aq)$ ions, and $CO_2(aq)$ is formed which can be removed by transport to the lungs and then breathed out as $CO_2(g)$.

  - If $[H^+(aq)]$ falls (and the pH rises), then the equilibrium shifts to the right in order to raise the $[H^+(aq)]$ again. Thus the constant value of the equilibrium constant is maintained.

- The proteins present in blood, do themselves act as buffers because they have side groups such as $-COO^-$ and $-NH_3^+$ that can accept or donate protons, thus raising or lowering the pH.

- Hydrogen phosphate ions (HPO_4^{2-}) can buffer the solution around neutral pH by the reaction:

  $$\mathbf{HPO_4^{2-}(aq) + H_3O^+(aq) \rightleftharpoons H_2PO_4^-(aq) + H_2O(l)}$$

  Or more simply:

  $$\mathbf{HPO_4^{2-}(aq) + H^+(aq) \rightleftharpoons H_2PO_4^-(aq)}$$

✓ *Quick check 1*

Other uses of buffers

Situation/Problem	How the buffer is used
In scientific research to mimic conditions in the body	The buffer forms the solution in which the experiment is carried out. For example, phosphate buffer is used as a buffer for many reactions where the pH in the body is around neutral.
The urea in babies' urine can form ammmonia which is alkaline causing 'nappy rash' when bacteria act upon it.	Baby lotion is buffered at pH 6 thus lowering the pH and stopping the bacterial enzymes working.
The calibration of pH meters	Buffers provide a good constant calibration point for these meters.
Drugs and lotions for use in or on the body. For example, shampoos are buffered.	The pH of these mixtures must be kept at or near physiological pH in order that the body tissues are not adversely affected.
Industrial preparations	As a medium in which the pH is kept constant.
Washing powders use enzymes for 'cleaning' fabrics.	Buffers are included to help maintain a constant pH in which the enzymes work efficiently.

✓ Quick check 2

? *Quick check questions*

1. Explain the physiological importance of buffers.
2. Give some uses of buffers.

Indicators and acid–base titration curves

The theory of this section is an extension of the previous sections so make sure you understand terms like pH and pK_a.

Indicators

- Indicators are either weak acids or weak bases.
- The equilibrium for an indicator that is a weak acid may be written as:

$$\mathbf{HIn(aq) \rightleftharpoons H^+(aq) + In^-(aq)}$$

colour I   **colour II**

The pH range over which an indicator changes colour is equal to pK_a ± 1.

- Therefore, for HIn at a pH at least 1 pH unit below pK_a the colour of the indicator is colour I. At pH values at least 1 pH unit greater than pK_a it has the colour II.
- An indicator can be used for a titration if the pH range over which it changes colour coincides with a rapid pH change in the titration (see titration curves below).
- Two common indicators are phenolphthalein (abbreviated to PP; pK_a = 9.3) and methyl orange (abbreviated to MO; pK_a = 3.7). The diagram below shows the pH ranges of these two indicators.

pH 0 1 2 3 4 5 6 7 8 9 10 11 12 13 14

red ←→ yellow
methyl orange, pK_a = 3.7

colourless ←→ pink
phenolphthalein, pK_a = 9.3

✓ Quick check 1

Titration curves

For all the titrations below and on the next page, 0.1 mol dm^{-3} alkali (base) is added to 25 cm^3 of 0.1 mol dm^{-3} acid.

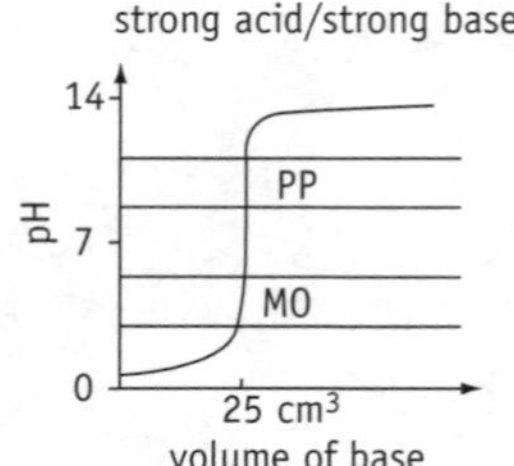

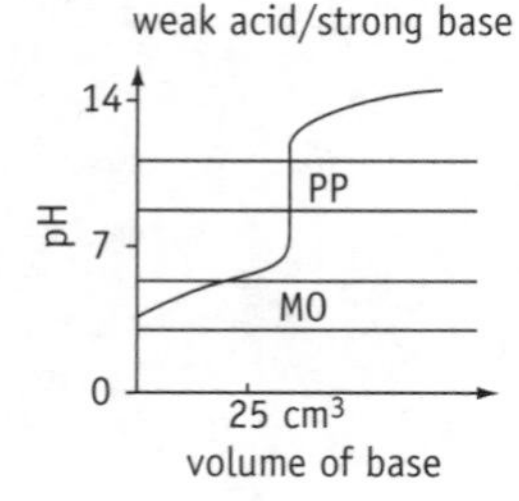

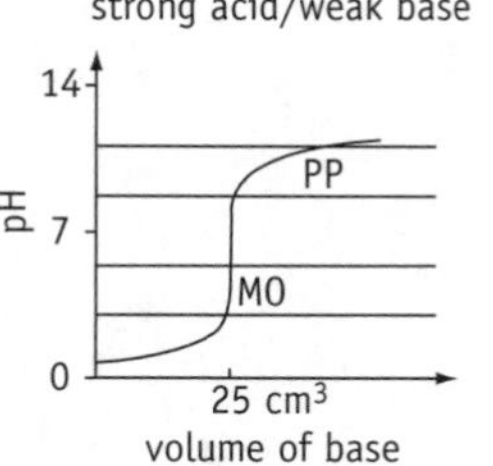

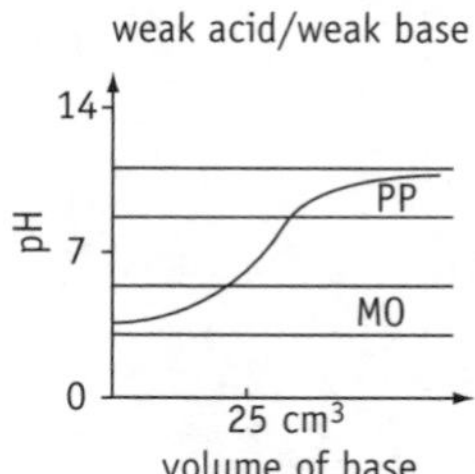

Choice of indicator

Titration	Which indicator	Reasons for choice
Strong acid/strong base	MO or PP	There is a rapid pH change over the colour range for either indicator.
Weak acid/strong base	PP	The rapid pH change is over the range for PP but not for MO.
Strong acid/weak base	MO	The rapid pH change is over the range for MO but not for PP.
Weak acid/weak base	Neither	There is no rapid pH change in any region of the titration curve.

✓ Quick check 2

? Quick check questions

The pK_a values of different indicators are given in the table below:

Indicator	pK_a	Colour change acid ⇌ alkaline
Congo red	4.0	violet ⇌ red
Methyl red	5.1	red ⇌ yellow
Thymol blue	8.9	yellow ⇌ blue
Thymolphthalein	9.7	Colourless ⇌ blue

1 What are the colours of the above indicators at the following pH values

(a) pH 1.0

(b) pH 7.0

(c) pH 10

2 Which of the above indicator(s) could you use for the following titrations:

(a) strong acid/strong base

(b) weak acid/strong base

(c) strong acid/weak base

(d) weak acid/weak base

Sample synoptic questions

(1) Copper readily forms complexes with water, ammonia and chloride ions, and these co-exist in equilibria.

(a) Concentrated hydrochloric acid is added to a solution containing $[Cu(H_2O)_6]^{2+}(aq)$ until there is a high concentration of Cl^-.

$$[Cu(H_2O)_6]^{2+}(aq) + 4Cl^-(aq) \rightleftharpoons [CuCl_4]^{-2}(aq) + 6H_2O(l)$$

hexaaquocopper(II) — **pale blue**; **tetrachlorocuprate(II)** — **yellow**

(i) Using one or both of the two complexes given above, explain what is meant by the following terms:

I dative covalent bonding [2]

II ligand [2]

III complex [2]

(ii) Draw the two complexes described in the equilibrium. [4]

(b) (i) State Le Chatelier's principle [2]

(ii) For the equilibrium shown, use Le Chatelier's principle to deduce what is observed as the acid is added. [3]

(c) The equilibrium constant, K_c, for the equilibrium in (a) may be written as:

$$K_c = \frac{\left[[CuCl_4]^{2-}\right]\left[H_2O\right]^6}{\left[[Cu(H_2O)_6]^{2+}\right]\left[Cl^-\right]^4}$$

It is possible in aqueous solution to simplify this expression to:

$$K_c' = \frac{\left[[CuCl_4]^{2-}\right]}{\left[[Cu(H_2O)_6]^{2+}\right]\left[Cl^-\right]^4}$$

The numerical value of K_c' for this equilibrium, at 25°C, is 4.17×10^5.

(i) What are the units of K_c'? [1]

(ii) An equilibrium mixture, at 25°C, contained 1.17×10^{-5} mol dm^{-3} $[Cu(H_2O)_6]^{2+}(aq)$ and 0.800 mol dm^{-3} Cl^-.

Calculate the equilibrium concentration of $[CuCl_4]^{2-}(aq)$. [2]

(iii) Suggest why, in aqueous solution, it is possible to simplify K_c to K_c'. [2]

(2)(a) (i) With the aid of examples, explain the meaning of the terms *strong acid* and *weak acid*. [5]

(ii) Describe and explain the use of buffer solutions. [5]

(b) Assuming the temperature to be 25°C, what is the pH of:

(i) 0.05 mol dm^{-3} hydrochloric acid [1]

(ii) 0.01 mol dm^{-3} sodium hydroxide? [2]

(c) Benzoic acid, $C_6H_5CO_2H$, is a weak acid with an acid dissociation constant, K_a, of 6.3×10^{-5} mol dm^{-3} at 25°C.

(i) Calculate the pH of 0.020 mol dm^{-3} benzoic acid at this temperature. [3]

(ii) Draw a sketch graph of the change in pH which occurs when 0.020 mol dm^{-3} potassium hydroxide is added to 25 cm^3 of 0.020 mol dm^{-3} benzoic acid until in excess. [3]

(iii) The pK_a values of some indicators are:

thymol blue	1.7
congo red	4.0
thymolphthalein	9.7

Which of these indicators would be most suitable for determining the end point of the titration between the benzoic acid and potassium hydroxide in (ii)? Explain your answer. [2]

(3) Methanol, CH_3OH, is used as an alternative fuel to petrol in racing cars. Although methanol is less volatile than petrol, its combustion in these engines is more complete.

(a) Hydrogen bonds help to explain why methanol is less volatile than petrol.

(i) Draw a diagram to show hydrogen bonding between two molecules of methanol. Include any relevant dipoles in your diagram. [2]

(ii) What bonds exist between the molecules in petrol? [1]

(iii) Suggest the advantage of the combustion of methanol being more complete than that of petrol. [1]

(b) Enthalpy changes can be determined indirectly using standard enthalpy changes of formation.

Calculate the standard enthalpy change of combustion of methanol using the following data.

Compound	**ΔH_f /kJ mol^{-1}**
$CH_3OH(l)$	−238.9
$CO_2(g)$	−393.7
$H_2O(l)$	−285.9

[3]

(c) Despite the combustion of methanol being a highly exothermic reaction, methanol does not spontaneously burst into flame and requires a spark for ignition.

Using a Boltzmann distribution explain why this should be so. [4]

One of the consequences of the heat generated by a car engine is the formation of nitrogen oxide, NO.

$O_2(g) + 2NO(g) \rightleftharpoons 2NO_2(g)$

(i) Explain, in terms of changes in oxidation number, why this reaction is considered to be a redox reaction. [3]

In an experiment to investigate the effects of changing concentrations on the rate of reaction, the following results were obtained.

Experiment number	**Initial concentration $O_2/10^{-2}$ mol dm^{-3}**	**Initial concentration $NO/10^{-2}$ mol dm^{-3}**	**Initial rate of disappearance of $NO/10^{-4}$ mol dm^{-3} s^{-1}**
1	1.0	1.0	0.7
2	1.0	2.0	2.8
3	1.0	3.0	6.3
4	2.0	2.0	5.6
5	3.0	3.0	18.9

(ii) Deduce the order of reaction with respect to O_2 and NO, explaining your reasoning. [4]

(iii) I Write the rate equation for this reaction.

II Calculate the value of the rate constant, *k*. State its units. [3]

Answers to quick check questions

Unit D: Chains Rings and Spectroscopy

Arenes Page 3

1 (a) Yes, (b) No (c) Yes (d) No

2 (a) to (e)

(a)

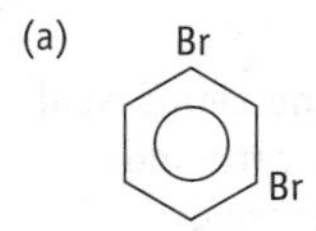

1,3-dibromobenzene

(b)

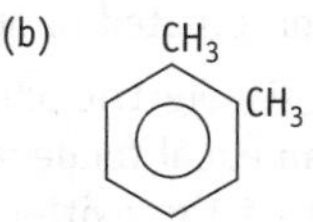

1,2-dimethylbenzene

(c)

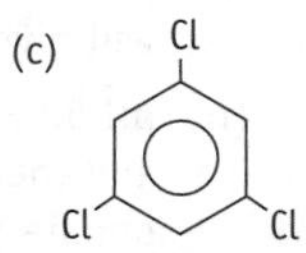

1,3,5-trichlorobenzene

(d)

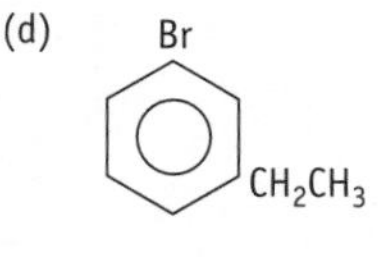

bromo-3-ethylbenzene

(e)

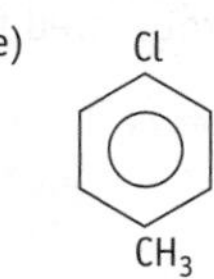

chloro-4-methylbenzene

3

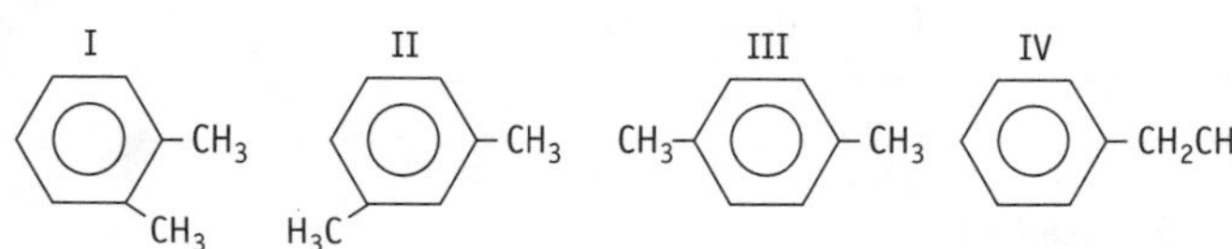

4

$C_6H_6 + Cl_2 \xrightarrow[\text{room temp}]{AlCl_3} C_6H_5Cl + HCl$ electrophilic substitution

$CH_3CH_2CH{=}CH_2 + Cl_2 \xrightarrow[\text{(no catalyst)}]{\text{room temp}} CH_3CH_2CH(Cl){-}CH_2(Cl)$ electrophilic addition

Electrophilic substitution Page 5

1 See text.

2 $C_6H_6 + HNO_3 \rightarrow C_6H_5NO_2 + H_2O$; For mechanism see text.

3 (a) **A** = C_6H_5Cl **B** = HCl
conditions; catalyst (halogen carrier) at room temperature

(b) **C** = CH_3Cl **D** = HCl
conditions; catalyst (halogen carrier) at 60 °C

(c) **E** = $C_6H_5NO_2$ **F** = H_2O
conditions; conc. nitric acid, catalyst (conc. sulphuric acid) at 50 °C

(d) **G** = Br_2 **H** = C_6H_5Br
conditions; catalyst (halogen carrier) at room temperature

(e) **I** = C_2H_5Cl **J** = HCl
conditions; catalyst (halogen carrier) at 60°C

Phenol Page 7

1 See text.

2

(a) ⟶

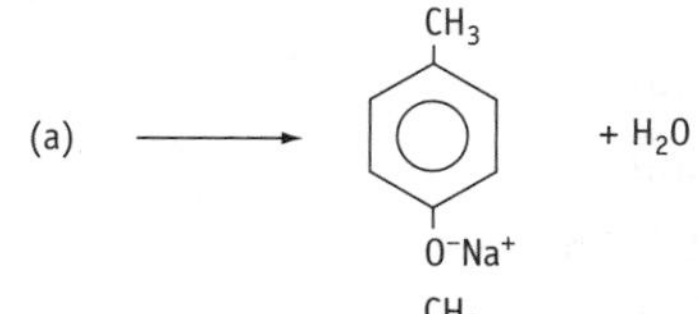

(b) ⟶ $CH_3C_6H_4O^-Na^+ + \frac{1}{2} H_2$

(c) ⟶ $BrC_6H_4O^-Na^+ + H_2O$

3 See text.

4 With phenol the reaction takes place with bromine water whilst benzene requires undiluted bromine liquid.
Phenol does not require a catalyst to react whilst benzene requires a halogen carrier (e.g. $AlCl_3$)

Carbonyl compounds Page 9

1 (a) butanal (b) 2-methylpropanal
(c) phenylmethanal (d) 2-phenylethanal

2 (a) propan-2-one (b) pentan-3-one
(c) 1-phenylpropan-2-one (d) 4-phenylbutan-2-one.

3 (a) $CH_3{-}C(OH)(CN){-}H$ (b) $CH_3{-}C(OH)(CN){-}CH_2CH_3$

4 (a) ethanol CH_3CH_2OH (b) butan-2-ol $CH_3CH_2CH(OH)CH_3$
(c) 1-phenylethanol $CH_3CH(C_6H_5)OH$
(d) 2-methylpropan-1-ol $CH_3CH(CH_3)CH_2OH$

Carboxylic acids and esters Page 11

1 (a) CH_3CH_2COOH (b) $CH_3CH(CH_3)COOH$ (c) $C_6H_5CH_2COOH$

2 (a) See text.

(b) 2-Chloroethanoic acid is the stronger acid because the chlorine will increase the attraction on the electrons in the O–H bond thus weakening this bond even more, so that the hydrogen is more easily lost making the molecule a stronger proton donor and hence a stronger acid.

3 (a) $CH_3COO^-Na^+ + H_2O$
(b) $CH_3CHClCOO^- + H_2O$
(c) $C_6H_5COO^- + H_2O$
(d) $CH_3CH(CH_3)COO^-Na^+ + H_2O$

4 Both acids are catalysts but concentrated sulphuric also moves the equilibrium to the right-hand side (and more ester). See text.

Esters Page 13

1 (a) CH_3CH_2COOH CH_3COOCH_3 $HCOOC_2H_5$

(b) $CH_3CH_2CH_2COOH$ $CH_3CH(CH_3)COOH$ $CH_3COOCH_2CH_3$ $HCOOCH_2CH_2CH_3$ $HCOOCH(CH_3)CH_3$

2 (a)(i) propanoic acid, CH_3CH_2COOH (ii) methanoic acid, HCOOH (iii) butanoic acid $CH_3CH_2CH_2COOH$

(b)(i) methyl propanoate (ii) methyl methanoate (iii) ethyl butanoate

3 See text.

4 (a) CH_3COOH and C_2H_5OH

(b) C_6H_5COOH and CH_3OH

(c) CH_3CH_2COOH and C_6H_5OH

(d) HCOOH and $CH_3CH_2CH_2OH$

Nitrogen compounds Page 15

1 (a) methylamine (b) propylamine
(c) 2-methylpropylamine (d) phenylamine

2 (a) (i) $CH_3NH_3^+ + OH^-$ (ii) $C_6H_5NH_3^+ + OH^-$

(b) They can **accept protons** because the lone-pair electrons on the nitrogen can form a dative covalent bond with the proton.

3 (a) $C_6H_5NH_2 < NH_3 < CH_3NH_2$. See text for explanation.

(b) It would be **less** basic because the electron density on the nitrogen is reduced by the electron-withdrawing $-NO_2$ group on the ring making it less able to accept a proton.

4 (a) $CH_3NH_3^+Cl^-$ name = methylammonium chloride

(b) $CH_3CH_2CH_2NH_3^+Cl^-$ name = propylammonium chloride

(c) $C_6H_5NH_3^+Cl^-$ name = phenylammonium chloride

Phenylamine and amino acids Page 17

1 (a) i $C_6H_5NH_2 + 2H_2O$ name = phenylamine
ii $C_6H_4(CH_3)NH_2 + 2H_2O$ name = 4-methylphenylamine

(b) Tin and concentrated hydrochloric acid.

2 (a) i **A** = $C_6H_5N_2^+\ Cl^-$ (benzene diazonium chloride)

B = $C_6H_5{-}N{=}N{-}C_6H_4{-}OH$

(ii) **C** = $C_6H_4(CH_3)N_2^+Cl^-$

D = $CH_3C_6H_4{-}N{=}N{-}C_6H_3(CH_3){-}OH$

(b) They are chemical dyes.

3 (a) (i)
$CH_2(NH_3^+)COOH\ Cl^-$ (ii) $CH_2(NH_2)COO^-Na^+ + H_2O$

(b) In (i) it accepts a proton from HCl and in (ii) it donates a proton to OH^-.

Amino acids (zwitterions), proteins and peptides Page 19

1

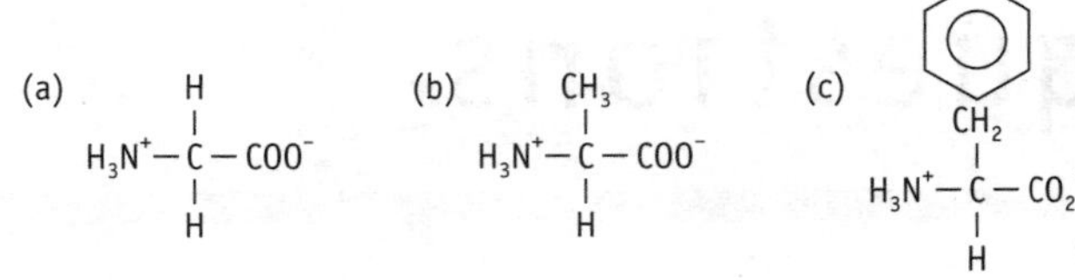

2

(a) (i) $H_3N^+{-}CH_2{-}COOH$ (ii) $H_3N^+{-}CH_2{-}COO^-$ (iii) $H_2N{-}CH_2{-}COO^-$

Explanations

(i) pH 2.00 is below the pI of the amino acid and it will give the protonated form.

(ii) pH 5.97 is the isoelectric point of the amino acid and there is an equal tendency to accept and donate protons – the zwitterion is formed.

(iii) pH 10.00 is above the pI of the amino acid and it will give the deprotonated form.

(b) (i) $H_3N^+{-}CH(CH_3){-}COOH$

(ii) $C_6H_5CH_2{-}CH(NH_2){-}COO^-$

3 See text.

4 (a) (i)

$H_2N{-}CH_2{-}CO{-}NH{-}CH(CH_3){-}COOH$ and $H_2N{-}CH(CH_3){-}CO{-}NH{-}CH_2{-}COOH$

dipeptide I dipeptide II

(ii)

$H_2N{-}CH_2{-}CO{-}NH{-}CH(CH_2C_6H_5){-}COOH$ and $H_2N{-}CH(CH_2C_6H_5){-}CO{-}NH{-}CH_2{-}COOH$

dipeptide I dipeptide II

(iii) $H_2N{-}CH_2{-}CO{-}NH{-}CH_2{-}COOH$

(b) (i)

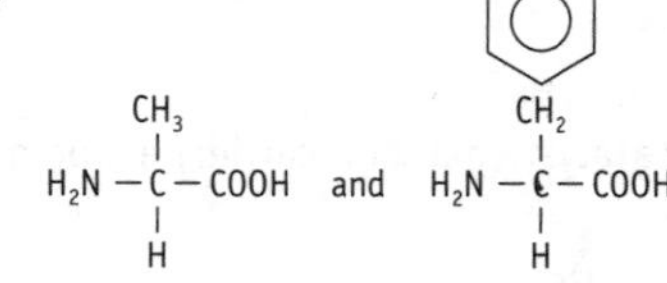

(ii)

```
        CH3                        H
        |                          |
H2N — C — COOH  and  H2N — C — COOH
        |                          |
        H                     H — C — CH3
                                   |
                                   CH3
```

Stereoisomerism and organic synthesis Page 21

1 (a) (i) See text. (ii)

```
C6H5      CH3        C6H5       H
    \    /               \     /
     C = C                C = C
    /    \               /     \
   H      H             H       CH3

  cis isomer           trans isomer
```

(b) Has two identical groups/atoms on one or both of the carbons in the C=C bond.

2 (a)

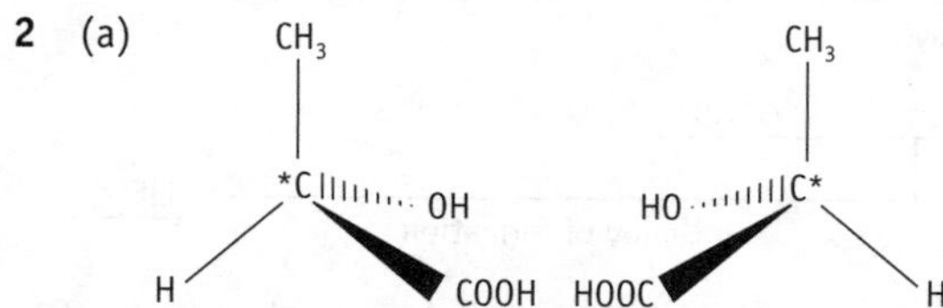

(b) In $CH_3C^*(CH_3)(OH)COOH$ the carbon marked with an asterisk does not have four different groups attached to it and therefore is not a chiral carbon atom.

3 The second step of the synthesis involves nucleophilic attack by CN^- on the carbonyl carbon. The ethanal is planar and therefore the attack can be from above or below leading to equal numbers of the possible isomers. See text for diagrams.

Chemical synthesis and synthetic pathways Page 23

1 See text.

2 See text.

3 (a) Pass $CH_2{=}CH_2$ and steam over heated phosphoric acid catalyst at a high pressure. Reflux the CH_3CH_2OH with excess acidified dichromate. React the CH_3COOH with ethanol in the presence of concentrated sulphuric acid catalyst.

(b) React CH_3OH with HBr (NaBr + conc. H_2SO_4) React CH_3Br with KCN in alcohol and gentle reflux. Reflux CH_3CN with aqueous HCl.

(c) React $CH_3CH_2CH_2CHO$ with KCN in the presence of a low concentration of acid at room temperature. Reflux $CH_3CH_2CH_2CH(OH)CN$ with aqueous HCl. React the $CH_3CH_2CH_2CH(OH)COOH$ with NaBr and conc. H_2SO_4 (HBr).

4 **A** is $CH_3CH(OH)CN$ **B** is $CH_3CH(OH)COOH$ **C** is $CH_3CHBrCOOH$ **D** is $CH_3CH(NH_2)COOH$

Polymerisation Page 25

1

```
   (a)             (b)             (c)            (d)
 [ CN  CN ]     [ CH3  H ]     [ H   Cl ]     [ F   F ]
 [ |   |  ]     [ |    | ]     [ |   |  ]     [ |   | ]
-[ C — C  ]-   -[ C  — C ]-   -[ C — C  ]-   -[ C — C ]-
 [ |   |  ]     [ |    | ]     [ |   |  ]     [ |   | ]
 [ CN  CN ]n    [ CH3  H ]n    [ H   H  ]n    [ F   F ]n
```

2 See text.

3

(a) $\left[\!-OCH_2CH_2OC(=O)-C(=O)-\right]$

(b) $\left[\!-OCH_2CH_2OC(=O)-C_6H_4-C(=O)-\right]$

Condensation polymerisation Page 27

1

(a) $\left[\!-C(=O)-CH_2CH_2C(=O)-O-CH_2CH_2O-\right]$

(b) $\left[\!-C(=O)(CH_2)_4C(=O)-O(CH_2)_6O-\right]$

(c) $\left[\!-C(=O)-C_6H_4-C(=O)-NH-C_6H_4-NH-\right]$

2 (a) $HOOCCH_2CH_2COOH$ and $HOCH_2CH_2OH$

(b) $HOOC(CH_2)_4COOH$ and $HO(CH_2)_6OH$

(c) $HOOC-C_6H_4-COOH$ and $H_2N-C_6H_4-NH_2$

3 See text.

Infrared spectroscopy Page 29

1 (a) $CH_3CH_2CH_2COOH$

(b) The absorption at ~1700 cm^{-1} is due to the >C=O group (stretch) whilst the broad absorption at ~3000 cm^{-1} is due to the presence of an –OH group thus showing that there is a –COOH group present.

2 (a) –OH

(b) alcohol

Not –COOH because this would also give an absorption at ~1700 cm^{-1} due to the >C=O group.

3 At first there would be an absorption at 3000 cm^{-1} due to the –OH group and then this would gradually disappear and an absorption peak at ~1700 cm^{-1} would appear due to the formation of >C=O in CH_3CHO.

NMR Page 31

1 (a) Four chemically different types of proton are present in the molecule.

(b) There are two chemically different protons on a carbon adjacent to that carbon with the protons responsible for the peak.

2 (a)

```
    H  H            H  H
    |  |            |  |
 H— C— C—H      Cl— C— C—H
    |  |            |  |
    Cl Cl           Cl H
```

1,2-dichloroethane 1,1-dichloroethane

(b) 1,2-dichloroethane

(c) There is one peak and therefore only one type of proton. 1,2-dichloroethane has one type of proton whilst 1,1-dichloroethane has two types.

Identifying compounds using NMR Page 33

1 (a)

```
  H H  O          H O H
  | |  //         | ‖ |
H-C-C-C         H-C-C-C-H
  | |  \          |   |
  H H   H         H   H
 propanal     propan-2-one
```

(b) Propanal There are three peaks and therefore three types of proton. Propanal has this number of proton types, propan-2-one has only type. Therefore the compound must be propanal.

(c) Explanation

The peak at $\delta = \sim 10$ ppm is due to the aldehyde proton on the propanal.

The quadruplet at $\delta = \sim 2.5$ ppm is due to the -**CH_2** which are split by the adjacent three -**CH_3** protons into n+1, i.e. 4 peaks.

The triplet at $\delta = \sim 1$ ppm is due to the presence of the three -**CH_3** protons which are split by the adjacent two -**CH_2** protons.

2 (a)

```
  H  O
  |  //
H-C-C
  |  \
  H   O-H
```

(b) Two types of proton.

(c) The carboxyl (COO**H**) proton at $\delta = \sim 11.7$ ppm and the -**CH_3** protons at $\delta = \sim 2$ ppm.

(d) The peak at $\delta = \sim 11.7$ ppm will disappear because the carboxyl proton will exchange with the D_2O to give -COO**D** which will not produce a peak in this energy region.

Mass spectroscopy Page 33

1 72

2 The molecular mass of benzoic acid, C_6H_5COOH is 122 and there is a molecular ion peak at m/e = 122.

Unit E: Trends and Patterns

Lattice enthalpy Page 39

1 NaCl has the more exothermic lattice enthalpy, because Na^+ is smaller than Cs^+ so has a higher charge density. Therefore it has a higher electrostatic attraction to Cl^- in the lattice.

2 $Al^{3+}(g) + 3Cl^-(g) \rightarrow AlCl_3(s)$

3 Magnesium oxide has a large exothermic lattice enthalpy so has a very high melting point, which makes it suitable for use in furnaces.

The Born–Haber cycle Page 41

1

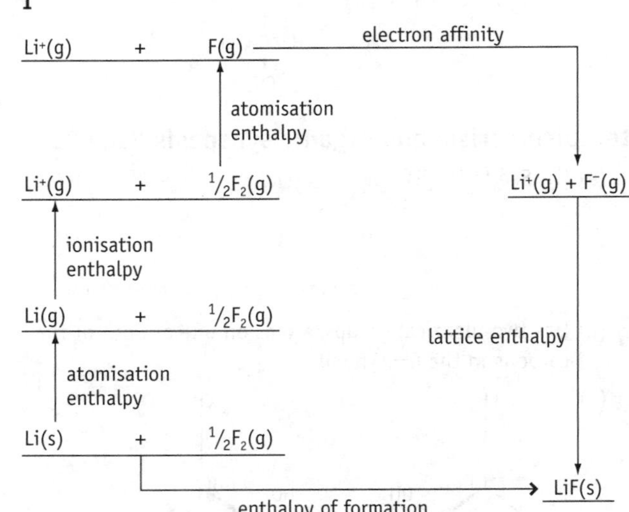

Things to know about Born–Haber cycles Page 43

1 $\Delta H_f\ (CaCl_2) = (+178) + (+590) + (+1145) + (+244) + (-698) + (-2258)$

$= -799\ kJ\ mol^{-1}$

The decomposition of the Group 2 carbonates Page 45

1 The decomposition temperatures of the Group 2 carbonates become higher going down the Group.

$MgCO_3(s) \rightarrow MgO(s) + CO_2(g)$

2 The lattice enthalpies of the Group 2 carbonates become less exothermic going down the Group.

3 The magnesium cation is smaller than the barium cation so has a higher charge density. Therefore it polarises the carbonate anion to a greater extent, which makes the regular lattice distorted.

4 Magnesium oxide has a stronger lattice than calcium oxide because the lattice enthalpy of MgO is more exothermic than the lattice enthalpy of CaO. This is because Mg^{2+} is smaller than Ca^{2+}.

Reactions of some Period 3 elements with water and with oxygen Page 47

1 (a) $Mg(s) + 2H_2O(l) \rightarrow Mg(OH)_2(aq) + H_2(g)$

$Mg(s) + H_2O(g) \rightarrow MgO(s) + H_2(g)$

2 $4Al(s) + 3O_2(g) \rightarrow 2Al_2O_3(s)$

The product, aluminium oxide, is a white solid. It does not react with water.

3 The trend in the structure of the oxides is from giant lattice to simple molecular going from left to right across the Period. The trend in the bonding is from ionic to covalent in the same direction. For examples see text.

4 The solution is acidic, because phosphorus is a non-metal on the right of the Period, so the oxide gives an acidic solution when dissolved in water.

Reactions of some Period 3 elements with chlorine Page 49

1 PCl_3 is a liquid with simple covalent molecules. PCl_5 is a solid with a giant ionic lattice composed of PCl_4^+ and PCl_6^- ions. The bonding within the ions is covalent.

2 The reaction of $SiCl_4$ with water is a hydrolysis reaction. It is exothermic, gives off white fumes and gives a white solid and an acidic solution.

$SiCl_4(l) + 4H_2O(l) \rightarrow Si(OH)_2(s) + 4HCl(g)$

3 $Mg(s) + Cl_2(g) \rightarrow MgCl_2(s)$

Magnesium chloride has a giant lattice structure with ionic bonding. It dissolves in water to give a very slightly acidic solution, pH 6.5

$PCl_3(l) + 3H_2O(l) \rightarrow H_3PO_3 + 3HCl(g)$ pH = 2

Transition elements – electronic configurations Page 51

1 A transition element has at least one ion with a partially filled d sub-shell.

2 Mn: [Ar] $3d^5$ $4s^2$; Cr: [Ar] $3d^5$ $4s^1$

3 Mn^{2+}: [Ar] $3d^5$; Cr^{3+}:[Ar] $3d^3$

Transition metals – oxidation states, catalytic behaviour and the hydroxides Page 53

1 Transition metals have variable oxidation states because the electrons in the d sub-shell have very similar energies. Examples are iron with common oxidation states of +2 and +3, and copper with oxidation states of +1 and +2.

2 Finely divided to give a large surface area; iron is a transition metal so transfers electrons easily; iron also provides a solid site for the reaction to take place.

3 Green precipitate with $OH^-(aq)$ so the metal is iron.

$Fe^{2+}(aq) + 2OH^-(aq) \rightarrow Fe(OH)_2(s)$

Transition elements and colour Page 55

1 CuCl contains Cu^+, which has an electronic configuration of [Ar] $3d^{10}$. Colour only occurs with a partially filled 3d sub-shell, not a full 3d sub-shell, so aqueous CuCl is colourless.

2 The colour of the solution is green. The solution absorbs at 650 nm which is red light so the colour seen is the colour opposite red in the colour wheel – green.

Complex ions Page 57

1 A ligand is a molecule or ion which forms a dative covalent bond to a central metal atom or ion in a complex ion. For examples, see text.

2 (a) octahedral (b) octahedral (c) tetrahedral

3 $[Cu(H_2O)_6]^{2+}(aq) + 4Cl^-(aq) \rightarrow [Cu(Cl)_4]^{2-}(aq) + 6H_2O(l)$

$[Cu(Cl)_4]^{2-}(aq) + 4NH_3(aq) + 2H_2O(l) \rightarrow [Cu(NH_3)_4(H_2O)_2]^{2+}(aq) + 4Cl^-(aq)$

As the concentrated hydrochloric acid is added the solution changes from blue to yellow. When the concentrated ammonia solution is added the solution turns blue again and then changes to deep blue.

Colorimetry and the formulae of complex ions Page 59

1 The tube with the maximum absorbance is tube 6, which contains equal number of moles of Ni^{2+} and edta. So the formula is $[Ni(edta)]^{2+}$.

Redox reactions and titration calculations Page 59

1 $H_2O_2(aq) + 2H^+(aq) + 2e^- \rightarrow 2H_2O(l)$

2 $2Fe^{2+} \rightarrow 2Fe^{3+} + 2e^-$ (×2 to make the numbers of electrons in each equation the same)

add together and cancel the electrons on each side:

$H_2O_2(aq) + 2H^+(aq) + 2Fe^{2+} \rightarrow 2Fe^{3+} + 2H_2O(l)$

2 no. moles MnO_4^- = 14.9/1000 × 0.02 = 2.98×10^{-4} mol

$MnO_4^-(aq) + 5Fe^{2+}(aq) + 8H^+(aq) \rightarrow Mn^{2+}(aq) + 5Fe^{3+}(aq) + 4H_2O(l)$

∴ no. moles Fe^{2+} = $2.98 \times 10^{-4} \times 5 = 1.49 \times 10^{-3}$ mol

∴ concentration of $Fe^{2+}(aq)$ = $(1.49 \times 10^{-3}) \div (100/1000)$ = 0.0149 mol dm^{-3}

Unit F: How Far, How Fast?

How Fast? The rate equation Page 65

1 (a) first order with respect to NO
(b) second order overall

2 $dm^3\ mol^{-1}\ s^{-1}$

How Fast? Calculating *k*, the effect of *T* and reaction mechanisms Page 67

1 $3.00 \times 10^{-4}\ s^{-1}$

2 The rate-determining step is the reaction between a CH_3COCH_3 molecule and an H^+ ion.

How Fast? Concentration–time graphs Page 69

1 Yes, it has a constant half-life.

How Fast? Rate–concentration graphs Page 71

1 Measure the initial rate of a reaction in several experiments in which the reactants have different concentrations. For graph see text.

2 A graph of rate against concentration is a straight line with a gradient, so this is a first-order reaction.

How Far? The equilibrium law and K_c Page 73

1 (a) $K_c = \frac{[PCl_3][Cl_2]}{[PCl_5]}$ mol dm^{-3}

(b) $K_c = \frac{[C_2H_5OH]}{[C_2H_4]\ [H_2O]}$ $dm^3\ mol^{-1}$

(c) $K_c = \frac{[HBr]^2}{[H_2][Br_2]}$ no units

How Far? How to calculate the value of K_c Page 75

1 $K_c = 0.27$ no units

How Far? The equilibrium law and K_p Page 77

1 $K_p = 0.14$ kPa

Unit F: Unifying concepts

Acids and bases Page 79

1 (a) $HClO_4$ and ring ClO_4^- (b) HBr and ring Br^-
(c) H_2O and ring OH^-

2 (a) hydroxonium or oxonium ion (b) In 1(b) the water accepts a proton (H^+ ion) from HBr and is therefore a base. In 1(c) it donates a proton to the CH_3NH_2 and therefore acts as an acid. Because it can act as both an acid and a base it is amphoteric.

3 (a) $HClO_4 + H_2O \rightleftharpoons ClO_4^- + H_3O^+$
$HNO_3 + H_2O \rightleftharpoons NO_3^- + H_3O^+$
(b) The $HClO_4$ is the stronger proton donor because it is the acid in its reaction with HNO_3.

Acids and bases in aqueous solution Page 81

1 (a) 2 (b) 8.7 (c) 3.52

2 (a) 3.16×10^{-3} mol dm^{-3} (b) 3.98×10^{-8} mol dm^{-3}
(c) 3.16×10^{-11} mol dm^{-3} (d) 2.51×10^{-14} mol dm^{-3}

3 (a) (i) 2 (ii) 4.52
(b) (i) 12 (ii) 9.48

The chemistry of weak acids Page 83

1 (a) pK_a (HA) = 2.54 pK_a (HB) = 5.35
(b) HA is the stronger acid because it has the lower pK_a (or conversely the higher K_a).
(c) $HA + HB \rightleftharpoons A^- + H_2B^+$

2 $[H^+(aq)] = 10^{-pH} = 10^{-3.41} = 3.89 \times 10^{-4}$
$K_a = [H^+(aq)]^2/0.01 = (3.89 \times 10^{-4})^2/0.01$
$= 1.51 \times 10^{-5}$ mol dm^{-3}

3 (a) $K_a = 10^{-pK_a} = 10^{-9.9} = 1.26 \times 10^{-10}$ mol dm^{-3}
(b) $[H^+(aq)] = \sqrt{(1.26 \times 10^{-10} \times 0.001)}$
$= 3.55 \times 10^{-7}$
pH $= -\log_{10}[H^+(aq)] = 6.44$

Buffers – how they work and their pH Page 85

1 (a) See text.

2 (a) [HF] = 100/150 × 0.03 = 0.02 mol dm^{-3}
(b) [NaF] = 50/150 × 0.03 = 0.01 mol dm^{-3}
(c) $[H^+(aq)] = K_a \times [HF]/[NaF] = 5.6 \times 10^{-4} \times 0.02/0.01$
$= 1.12 \times 10^{-3}$ mol dm^{-3}
pH $= -\log_{10}[H^+(aq)] = -\log_{10}1.12 \times 10^{-3} = 2.95$
(d) The equilibrium for the acid is
$HF(aq) \rightleftharpoons H^+(aq) + F^-(aq)$

If H^+(aq) ions are added to disturb the equilibrium, they combine with the conjugate base, the F^-(aq) ions from the NaF, to give the largely undissociated HF acid. This removes them from the solution and the pH remains virtually constant. If OH^-(aq) ions are added they are neutralised by the H^+(aq) ions. This disturbs the equilibrium and more HF dissociates so that the value of K_a is restored, along with the concentration of H^+(aq) ions and the pH returns to its original value.

The importance of buffers in the body Page 87

1 See text.

2 See text.

Indicators and acid–base titration curves Page 89

1

Indicator	**a** Colour at pH 1.0	**b** Colour at pH 7.0	**c** Colour at pH 10.0
Congo red	violet	red	red
Methyl red	red	yellow	yellow
Thymol blue	yellow	yellow	blue
Thymolphthalein	colourless	colourless	colourless with perhaps a tinge of blue.

2 (a) Any of the four can be used.
(b) thymol blue or thymolphthalein
(c) congo red or methyl red
(d) No indicator can be used.

Answers to end-of-unit questions

NOTE * indicates a mark awarded.

Unit D: Chains, Rings and Spectroscopy

1 (a) (i) Let there be 100 g of the hydrocarbon

Element	%	Mass in 100 g	No. of moles	Relative no. of atoms
Carbon	90.56	90.56 g	90.56/12 = 7.55	4
Hydrogen	9.44	9.44 g	9.44/1 = 9.44	5 (because 9.44/7.55 =1.25 =5/4) * *

(ii) The empirical formula mass = (4 x 12) + (5 x 1) = 53

Therefore molecular formula = 2 x empirical formula = 2 x C_4H_5 = C_8H_{10} *

(iii)

CH_3 * CH_3 (benzene ring, 1,2-positions) CH_3 * CH_3 (benzene ring, 1,3-positions) CH_3 * CH_3 (benzene ring, 1,4-positions) CH_2CH_3 * (benzene ring)

(b) (i) A must be the hydrocarbon with the methyl groups at the 1 and 4 positions. The product can only have one identity. The others give monochloro-products which have the chlorine in more than one possible position .* (Draw them to prove it to yourself.)

(ii) B = benzene ring with CH_3, Cl, CH_3 *

2 (a) To form optical isomers, a compound should possess a chiral carbon (one with four different groups attached to it). Glycine does not possess this property because it has two hydrogens attached to the central carbon atom.

(b)

* $H_2N-C(CH_2C_6H_5)(H)-CO_2H + HCl \longrightarrow H_3N^+-C(CH_2C_6H_5)(H)-CO_2H + Cl^-$

* $H_2N-C(CH_3)(H)-CO_2H + NaOH \longrightarrow H_2N-C(CH_3)(H)-COO^- + Na^+ + H_2O$

(c) (i)

* $H_3N^+-C(H)(H)-COO^-$

zwitterion for glycine at pH 5.97

(ii)

$H_3N^+-C(CH_3)(H)-COOH$ * $H_2N-C(CH_2C_6H_5)(H)-COO^-$ *

alanine at pH 5.7 phenylalanine at pH 5.7

You are not asked for an explanation but the reasoning is that pH 5.75 is below the isoelectric point of alanine, so H^+ ions can be added to the zwitterion giving a protonated carboxylate group.

pH 5.75 is above the isoelectric point of phenylalanine so H^+ ions are removed giving a de-protonated zwitterion.

(d)

$HOOC-C(H)(H)-N(H)-C(=O)-C(CH_3)(H)-N(H)-H$ *

3 (a)

$C_6H_5-C(H)(H)-C(CH_3)(H)-\ddot{N}H_2$ * — lone pair electrons

(b) (i)

$C_6H_5-NH_2$ * $H-C(H)(H)-C(H)(H)-NH_2$ *

phenylamine ethylamine

(ii) It has a nitrogen atom with a lone pair of electrons which can accept a proton.*

(iii) Phenylamine has the nitrogen adjacent to the benzene ring and the lone-pair electrons can be delocalised on the ring* lowering the electron density on the nitrogen. This makes it harder for the nitrogen to accept a proton and the compound is therefore less basic than ethylamine, whose lone pair cannot be delocalised and are therefore readily donated to a proton.*

(iv) The nitrogen is not adjacent to the benzene ring and therefore the lone-pair electrons cannot be delocalised on the ring.*

The methyl group can push electrons towards the nitrogen increasing the electron density on the nitrogen making it more basic.*

4 (a) Compound P contains the >C=O group (absorption at ~1700 cm^{-1})* and the -OH group (broad absorption at ~3000 cm^{-1}).*

(b) (i) 4*

(ii) There are four distinct peaks (sets of peaks).*

(c)

$CH_3CH_2C(=O)-O-CH_3$ * — 3 types

$CH_3CH_2CH_2C(=O)OH$ * — 4 types

$HC(=O)-OCH_2CH_2CH_3$ * — 4 types

$HC(=O)-O-CH(CH_3)-CH_3$ * — 3 types

$CH_3-CH(CH_3)-C(=O)OH$ * — 3 types

(d) (i) P is butanoic acid because its infrared spectrum has two peaks- one at ~3000 cm^{-1} which is explained by the presence of the OH group and a peak at ~1700 cm^{-1} given by the >C=O group. The two groups make a carboxyl (COOH) group.*

(ii) The four peaks occur because there are four types of proton.*

The peak at $\delta = 11$ is the carboxyl proton.*

The peak at $\delta = 2.5$ is from the $-CH_2-$ group closest to the -COOH group. It is a triplet owing to the splitting effect of the two protons on the adjacent $-CH_2-$ group.*

The peak at $\delta = 1.7$ is produced by the next CH_2 group and it is a quadruplet because it is split by the three CH_3 protons.*

The peak at $\delta = 1$ is least affected by the -COOH group explaining its low δ value. It is a triplet because of the splitting by the two CH_2 protons adjacent to it.*

Unit E: Trends and Patterns

1 (a) Let there be 100 g of A.

	Fe	Cl
	44 g	56 g
no. of mol	44/55.8 = 0.789	56/35.5 = 1.57
relative number of atoms	1	2

A is $FeCl_2$ *

For B

	Fe	Cl
	2.79 g	8.12 – 2.79 = 5.33 g
no. of mol	2.79/55.8 = 0.05	5.33/35.5 = 0.15
relative number of atoms	1	3

B is $FeCl_3$ *

$Fe + 2HCl \rightarrow FeCl_2 + H_2$ *

$2Fe + 3Cl_2 \rightarrow 2FeCl_3$ *

(b) (i) C is hexaaquoiron(III) ion *$Fe(H_2O)_6^{3+}$ *

D is monothiocyanatopentaaquoiron(III) * $[Fe(H_2O)_5(SCN)]^{2+}$*

(ii) Water and SCN^- are ligands and they have lone-pair electrons. *

They form dative covalent bonds with the empty orbitals on the Fe^{3+} ion. *

The $Fe(H_2O)_6^{3+}$ is a complex ion with the central ion surrounded by six ligands; * diagram. *

(iii) When SCN^- is added to $Fe(H_2O)_6^{3+}$ it replaces one of the water molecules – one ligand is exchanged for another. *

(c) $FeCl_2$ (A) is ionic * – giant ionic structure, therefore high melting point *

$FeCl_3$ (B) is covalent * simple molecular * therefore lower melting point because less energy is required to separate molecules *

$FeCl_3$ is acidic because of hydrolysis *

$FeCl_3 + 3H_2O \rightarrow Fe(OH)_3 + 3HCl$ *

$FeCl_2$ simply dissociates (ionises) *

$FeCl_2 (aq) \rightarrow Fe^{2+} (aq) + 2Cl^- (aq)$ * (any 6)

2 (a) (i) and (ii)

All 6 correct in both rows *** Deduct 1 mark for each error.

Oxide	Na_2O	MgO	Al_2O_3	SiO_2	P_4O_{10}	SO_3
Melting point °C	1275	2827	2017	1607	580	33
Bonding	ionic	ionic	ionic	covalent	covalent	covalent
Structure	giant	giant	giant	simple molecular	simple molecular	simple molecular

(iii) MgO is a giant ionic structure therefore it has strong ionic bonds **throughout.** Therefore lots of strong bonds have to be broken and therefore high melting point *

SO_3 is covalent and is simple molecular with weak dipole–dipole forces between molecules which are easily broken *

(b) (i) $Na_2O + H_2O \rightarrow 2Na^+ + 2OH^-$ (or 2NaOH) *
pH 13–14 *

(ii) $SO_3 + H_2O \rightarrow 2H^+ + SO_4^{2-}$ (or H_2SO_4) * pH 1–2 *

(c) MgO would conduct because the ions are free to move and carry the current. *

SiO_2 would not conduct – it has no free ions or free electrons (no free charged particles) to carry the current. *

Synoptic exam questions

1 (a) (i) I A covalent bond (sharing of a pair of electrons) where both electrons come from one atom.*

In the examples, the oxygen in the water or the Cl^- provide both electrons in each dative covalent bond.*

II A ligand is a molecule or ion, in this case the water molecule or the Cl^- ion bonded via a dative covalent bond * to a central metal ion, in this case the Cu^{2+} ion. *

III The complexes in the example are formed by the Cu^{2+} ion accepting lone pair electrons into its empty orbitals * from the (ligands) water or Cl^- ion. *

(ii)

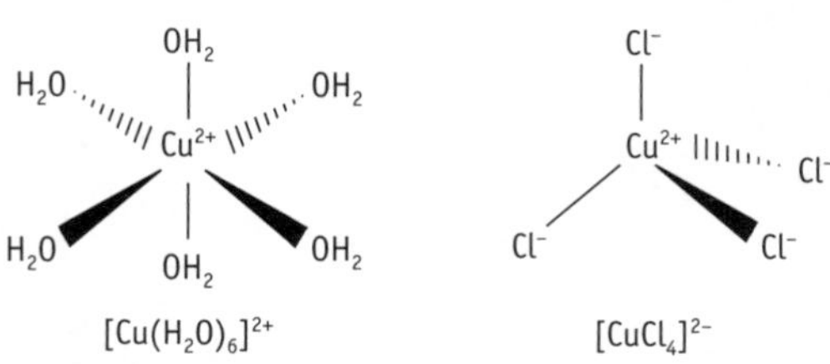

(b) (i) If changes are applied to a system in equilibrium * the equilibrium will move in the direction which tends to nullify * the effects of the change.

(ii) Solution changes to green and then to yellow * (these two must be the correct order to obtain the mark).

The equilibrium moves to the right * to reduce the concentration of Cl^-. *

(c) $dm^{12}\ mol^{-4}$ or $1/(mol^4\ dm^{-12})$ *

$4.17 \times 10^5 = x/1.17 \times 10^{-5} \times 0.8^4$ *

$x = 2.0\ mol\ dm^{-3}$ *

The concentration of water is very large * so is effectively constant. * Therefore the concentration of water can be incorporated into the equilibrium constant * (2 out of these 3)

2 (a) (i) an acid is a proton donor and a base is a proton acceptor *

The donation is complete if the acid is strong, and partial if the acid is weak. *

example of strong acid and weak acid *

example of strong base and weak base *

equation showing complete dissociation for strong acid OR base *

equation showing reversibility for weak acid OR base * (any 5)

(ii) A buffer solution resists change in pH when small quantities of acid or alkali are added *

suitable example e.g. weak acid and salt of weak acid *

how H^+ "mopped up" e.g. by reaction with anion or conjugate base*

equation for this reaction *

how OH^- "mopped up" e.g. by reaction with undissociated acid *

equation for this reaction *

two examples of use of buffer e.g. medicines, cosmetics, washing powders ** (any 5)

(b) (i) pH = 1.3 *

(ii) $[H^+] = 1 \times 10^{-14}/0.01$ *

pH = 12 *

(c) (i) $K_a = [H^+][C_6H_5COOH]/[C_6H_5COOH$ *

$6.3 \times 10^{-5} = [H^+]^2/0.02$ *

pH = 2.95 *

(ii) diagram to show:

shape of curve *

vertical section at 25 cm^3 *

graph starts at figure calculated in (i) or 2–4 and finishes at 12–14 *

(iii) thymolphthalein *

The pH must change quickly with a small addition of alkali OR pH change of indicator must be on vertical region of graph. *

3 (a) (i) correct bond shown between oxygen of one molecule and –OH hydrogen of other *

δ-"charges" shown on oxygen and hydrogen that are hydrogen bonded *

(if any wrong dipole charges are *shown, e.g. on C–H, then dipole mark is lost)*

The minimum required for 2 marks is shown below.

```
CH3— O
      \ δ+    δ−
       H - - - O — CH3
              /
             H
```

(ii) Any of the following will get the mark:

van der Waals / induced dipole–dipole / London / dispersion / instantaneous dipoles / temporary dipoles * (not "dipole–dipole" OR weak dipoles)

(iii) EITHER a "pollution mark": less or no carbon monoxide / carbon / soot / toxic gases OR fumes *

(not "not harmful")

OR an "energy mark": more heat / more energy / more exothermic *

(b) $CH_3OH(l) + 1\frac{1}{2}O_2(g) \rightarrow CO_2(g) + 2H_2O(l)$ *

One mark for getting the heat changes correct

–238.9, 0 –393.7 2 × –285.9 used (use of balancing numbers) *

Then equating the changes to the heat of reaction (i.e. correct cycle used)

(–393.7 + 2 × –285.9) – (–238.9) *

$-726.6\ kJ\ mol^{-1}$ * (correct numerical answer ignore units)

(c) Correct Boltzmann distribution with bell shaped curve * E_{act} indicated on diagram *

Explain that at room temperature (or no spark) very few methanol molecules have an energy greater than the E_{act} and therefore cannot react *

The spark provides the energy for most molecules to have an energy greater than E_{act} and therefore can react.

(i) $O_2 + 2NO \rightleftharpoons 2NO_2$

On L.H.S. the oxygen in O_2 has an oxidation state equal to 0. On the R.H.S. it is −2. *

Therefore it has been reduced. *

The nitrogen goes from oxidation state +2 to +4 on the R.H.S. *

Therefore it has been oxidised since its oxidation state has increased. *

Maximum 3 with italicised points for two of the three marks.

(ii) The order with respect to O_2 is first order *

Since using experiments 2 and 4 for example, doubling $[O_2]$ leads to a doubling of the rate. *

The order with respect to NO is second order *

Since using experiments 1 and 2 doubling [NO] causes the rate to quadruple *

(iii) (I) Rate = $k[O_2][NO]^2$

(II) Using experiment 1

$k = \text{Rate}/[O_2][NO]^2$

$= 0.7 \times 10^{-4}/1 \times 10^{-2} \times (1 \times 10^{-2})^2$ mol dm^{-3} s^{-1}/(mol dm^{-3})^3

= 70 * dm^6 mol^{-2} s^{-1} *

Index

Heinemann Educational Publishers
Halley Court, Jordan Hill, Oxford, OX2 8EJ
Part of Harcourt Education

Heinemann is a registered trademark of Harcourt Education Limited

First published 2001

10-digit ISBN: 0 435583 38 7
13-digit ISBN: 978 0 435583 38 5

08 07 06
10 9 8 7 6 5 4 3

Commissioning/development editor: Paddy Gannon

Edited by Peter Millward

Typeset by Saxon Graphics, Derby

Printed and bound in Great Britain by Ashford Colour Press Ltd, Gosport, Hampshire

Acknowledgements
The publishers have made every effort to trace the copyright holders, but if they have inadvertently overlooked any, they will be pleased to make the necessary arrangements at the first opportunity.

Tel: 01865 888058 www.heinemann.co.uk